SpringerBriefs in Earth Sciences

SpringerBriefs in Earth Sciences present concise summaries of cutting-edge research and practical applications in all research areas across earth sciences. It publishes peer-reviewed monographs under the editorial supervision of an international advisory board with the aim to publish 8 to 12 weeks after acceptance. Featuring compact volumes of 50 to 125 pages (approx. 20,000–70,000 words), the series covers a range of content from professional to academic such as:

- timely reports of state-of-the art analytical techniques
- bridges between new research results
- snapshots of hot and/or emerging topics
- literature reviews
- in-depth case studies

Briefs will be published as part of Springer's eBook collection, with millions of users worldwide. In addition, Briefs will be available for individual print and electronic purchase. Briefs are characterized by fast, global electronic dissemination, standard publishing contracts, easy-to-use manuscript preparation and formatting guidelines, and expedited production schedules.

Both solicited and unsolicited manuscripts are considered for publication in this series.

More information about this series at http://www.springer.com/series/8897

Chidozie Izuchukwu Princeton Dim

Facies Analysis
and Interpretation
in Southeastern Nigeria's
Inland Basins

 Springer

Chidozie Izuchukwu Princeton Dim
Department of Geology
Faculty of Physical Sciences
University of Nigeria, Nsukka
Nsukka, Enugu State, Nigeria

ISSN 2191-5369   ISSN 2191-5377   (electronic)
SpringerBriefs in Earth Sciences
ISBN 978-3-030-68187-6   ISBN 978-3-030-68188-3   (eBook)
https://doi.org/10.1007/978-3-030-68188-3

This Springer imprint is published by the registered company Springer Nature Switzerland AG
The registered company address is: Gewerbestrasse 11, 6330 Cham, Switzerland

All geoscience researchers who are interested in field geology and outcrop studies.

Acknowledgements

The author is most grateful to Emeritus Prof. K. M. Onuoha, Prof. A. W. Mode and Mr. I. C. Okwara of Department of Geology, University of Nigeria, Nsukka for immerse contributions that led to the publication of an aspect of this work. Thanks to the managements of Julius Berger Nigeria, Plc. Marlum Civil Engineering Nigeria Limited and Crush Stone Industry, for the access granted to their newly exposed quarry sites during field studies. Thanks also to Mr. L. M. Johnson, Dr. E. J. Adepehin, and Mr. E. C. Anigbogu for their assistance during various stages of geologic field visit.

Introduction

This research work focuses on utilizing information obtained from geologic outcrops and measured/logged stratigraphic sections in carrying out an outcrop-based facies analysis in the Cenomanian-Turonian succession of Eze-Aku Formation (Southern Benue Trough) and the Campanian succession of Afikpo Formation (lower intervals of Anambra Basin), across the Afikpo area (study area) in the southeastern part of Nigeria. Previous outcrop-based studies show that the Santonian Inversion that was associated with folding and magmatism affected the Cenomanian-Turonian statigraphic packages within the Afikpo area that gave rise to the tilted units and igneous emplacements which have been reported in the area. However, over the years, little or no good outcrops have hindered extensive and detailed study of the stratigraphic succession, which has led to limited understanding of the lithofacies and their distribution in the area. Recently, these rocks are being exposed through artisanal and large-scale quarrying activities. Hence, a call for a revisit that would allow for better understanding of the rock facies and the geology of the area. This work adopts an integrated approach that involves geological field studies and, facies and petrographic analysis in improving our understanding of outcropping lithofacies and their distribution, facies succession, facies association and environments of deposition in the Afikpo area of southeastern Nigeria. Emphasis will be on mapping of the lithostratigraphic units and identifying of lithofacies from outcropping rock successions, and recognizing facies succession from measured/logged stratigraphic sections. The target of this work is to classify these lithofacies into facies association, and reconstruct a facies depositional model that will improve the understanding of environments of occurrence of these sedimentary facies and the prevailing condition at their time of deposition. Overall, this work will demonstrate the application of outcrop-based facies analytical studies in interpreting the paleo-depositional environments of stratigraphic successions of frontier basins (such as those of the study area).

Keywords: Outcrop • Measured/Logged Stratigraphic Section • Facies Succession and Association • Paleo-Depositional Environment • Cretaceous successions • Abakaliki • Anambra Basins

Contents

Chapter 1
An Overview of Afikpo Study Area, Nigeria: Study Background, Reviews and Study Methodology

1.1 Introduction

Outcrop studies are commonly used to develop quantitative descriptions of sediment package and rock characteristics. Well log and seismic data provide excellent structural and stratigraphic information on subsurface, but may not be able to resolve small-scale vertical and lateral attributes of lithofacies. Outcrops can provide this important information provided they are of sufficient areal extent. Over the years, efforts at carrying out detailed studies of the Cenomanian–Campanian sediment packages within the ridge system of Southern Benue Trough and the lower intervals of the Anambra Basin, in Afikpo area of Southeastern Nigeria, which is the study area (Fig. 1.1), have been limited due to the lack of good outcrops.

Recently, quarrying activities through large-scale and artisanal mining have expose some good sections of the ridges, prompting a renewed interest in the area (Figs. 1.2 and 1.3). The present work was targeted at obtaining quantitative outcrop data through detailed geologic mapping and sedimentological logging across the existing outcrops and many of these newly exposed section around Afikpo town and environs. Hence, the need for a quantitative outcrop data collection has been driven by the requirement to improve the paleodepositional geological models needed for understanding of facies distribution in order improve their interpretation of environments of deposition.

1.2 Geographic, Geomorphic and Physiographic Setting

Geographically, the study area, which lies within the Southern Benue Trough, is located east of Afikpo town. It covers an area of approximately 574 km^2 and lies between latitudes 5° 53′ 48.47″ N and 5° 58′ 33.86″ N and longitude 7° 52′ 10.57″

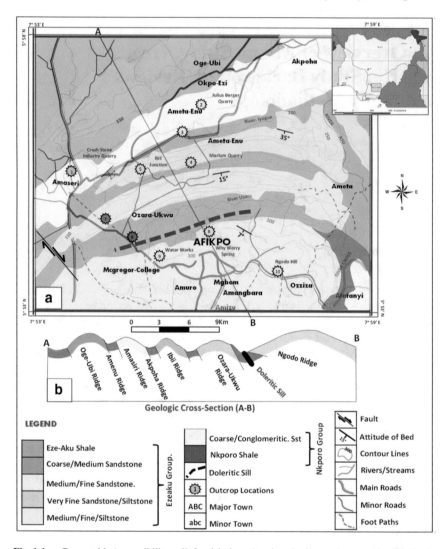

Fig. 1.1   a Geographic (accessibility, relief and drainage) and geologic map (outcropping lithologies and lithostratigraphic units) of the Afikpo area and environs, Southeastern Nigeria (*Note* contour is in feet). **b** Geologic cross-section (A–B), across the study area (after Dim et al. 2016)

E and 8° 00′ 36.04″ E (Fig. 1.1). The localities within this area include Okpo-Ezi, Ogo-Ubi and Akpoha in the North, Amate-Enu and Ibii, Central and Ameta, Amaseri, Ozara-Ukwu and Ngodo in the south, all in Ebonyi State, south-eastern Nigeria. The studied outcrops are accessible by all seasonal main roads, connecting Afikpo with Abakaliki (a distance of about 100 km). Accessibility to these areas are also possible by Okigwe - Abba Omega Road, Ugep - Abba Omega road,

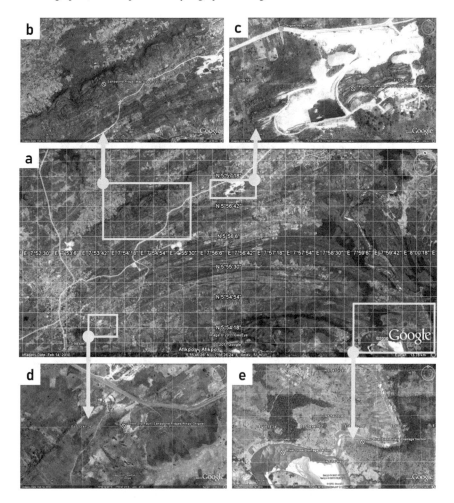

Fig. 1.2   a Satellite imagery map showing geographic, geomorphic and physiographic features across the study (Afikpo Area and Environs, Southeastern Nigeria). **b** Exposed section of the ridge through large-scale quarry section (Julius Berger, mining pit section) in some part of Akpoha area of Afikpo, along Abakaliki-Afikpo Road, Southeastern Nigeria. **c** Parallel to near-parallel continuous sandstone ridge stretching from eastern to western part in Ozara-Ukwu area of Afikpo area. **d** Structural distortion (strike slip fault) on Ozara-Ukwu Ridge at the southwestern part of the study area. **e** Drainage system showing the Cross River and Asu-River in the southeastern part of the study area. *Source* Google Earth™, 2010

Amagunze - Afikpo and Okpanku - Akaeze – Afikpo road. Most of the areas are connected with minor roads, main paths and minor paths (Fig. 1.1).

Geomorphologically, the study area shows an undulating rolling topography, which is controlled by the lithology and underlying bedrock. There is a gradual ascent from the plains in the southern part of the mapped area to the highland in the northern part of the study area formed by sandstones ridges. Elevation peaks

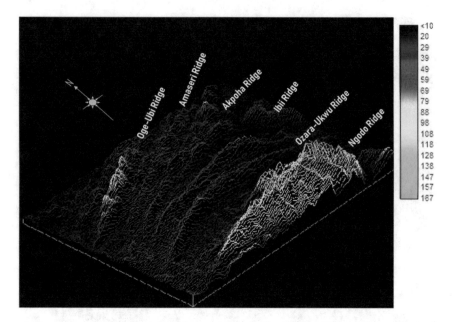

Fig. 1.3 3D Digital Elevation Map (DEM) of of the study area (Afikpo Area and Environs, Southeastern Nigeria) showing the sandstone ridges with the highest elevation (Ngodo Ridge) seen at the southern part of the study area (note: colour code bar is in meters) (after Dim et al. 2016)

of over 165 m are reached in sandstone ridges of Ogo-Ubi, Akpoha, Ibi, Amaseri, Ozara-Ukwu and Ngodo. These form the topographic highs (Figs. 1.1, 1.2 and 1.3). The physiographic features identified within the study area are the Cross River plains and Akpoha – Amaseri – Ibi - Ozalla-Ukwu – Afikpo sandstone ridges—these characterized its topography/relief and drainage systems (Figs. 1.2 and 1.3). Argillaceous sediments of Upper Cretaceous sequence underlie the plain. The plain extends beyond to the eastern axis of the study area and are between 60 m and 90 m above sea level. Satellite imagery shows that drainage is controlled by the Cross River and its tributaries, especially the Asu-River drainage system (Figs. 1.1 and 1.2). The 3Dimensional Digital Elevation Map (DEM) of the area reveals the major ridge systems (Fig. 1.3).

1.3 Review of Literatures

Many advances have been made in our understanding of facies interpretation and reconstructing their paleo-depositional environments (Miall 2000; Kendall 2005; Dim et al. 2016). The works of Miall (1996, 2000) show that in order to properly delineate lithofacies and understand its depositional environment, key attributes of the rock units such as rock type, textures and sedimentary structures must be taken into consideration. Several studies on lithofacies and their paleo-depositional environments have been carried within the stratigraphic successions of southeastern Nigerian sedimentary basins, which focus of this study.

Some of which are the work of Hoque (1977), which reported that the predominantly siliclastic Eze-Aku Formation (Turonian) located in the southeastern Nigerian Basin contains scattered lenses and layers of bioclastic limestone that show distinct facies of sparite-cemented grainstones, grainstones and wackestones within fine bioturbated silty sandstones and wackestone within laminated black shale. Wackestones are believed to be deposited by high-density turbidity currents carrying shallow-water shells into the deeper basin (Banerjee 1980). Simpsons (1954) first described the Eze-Aku Formation as comprising of hard, grey to black shale and siltstones that were deposited in a shallow marine environment. Banerjee (1980) developed a model for the various stages of the vertical growth in "Coarsening upward" successions typical of subtidal sand bars in the NE—trending linear sandstone bodies within marine shale of the Eze-Aku Formation (Upper Cretaceous) of southeastern Nigeria. Grainstones facies of Eze-Aku Formation (Turonian) has been interpreted as storm-lag deposits that either accumulated at the margins of offshore bars or on the wide platform of bioturbated muddy sand located below the wave base. Ichnological and lithological considerations suggested that Turonian calcareous sandstone sequence that forms part of the Eze-Aku Formation features an assemblage of ichnofossils in sediments that were deposited in an aerated shallow shelf environment. Deposition was generally below wave base under a continuous but relatively slow rate of sedimentation (Akpan and Nyong 1987). Palynological age and correlation of a black shale section in the Eze-Aku formation, Lower Benue trough, shows that the black, carbonaceous shales were probably deposited under anoxic bottom conditions marine environment (Lawal 1991).

Recently, efforts have been made by Ukaegbu and Akpabio (2009) and Okoro and Igwe (2014) to delineate lithofacies and interpret its depositional environments within the stratigraphic succession of the Eze-Aku Formation (Benue Trough) across the southeasten part of Nigeria. In the Maastrichtian coal bearing succession of the Anambra Basin of Nigeria, Dim et al. (2019), identify key lithofacies, classified them into several association and establish their depositional environments. On a regional scale (Simpsons 1954), Banerjee (1980) and Amajor (1980, 1987) have carried out some sedimentological studies within the Middle–Upper Cretaceous units of the southern Benue Trough. However, these works have been done using some limited or few outcrops sections considering the scale of their study area.

1.4 Aim and Objectives of Study

The aim of this study was to integrate outcrop-based data from several locations and information obtained from measured/logged lithologic sections with facies analysis to reconstruct paleo-depositional environment of sedimentary succession in the Afikpo area. In other words, this work demonstrates the application of outcrop studies and measured stratigraphic section interpretation for improved understanding of the lithofacies and their depositional environments across the folded/tilted Amaseri sandstone ridges, of the Turonian Eze-Aku Formation (Southern Benue Trough) and the Afikpo Formation (lower intervals of Anambra Basin).

The principle objectives of this research are;

- To map outcropping lithostratigraphic and rock units
- To identify lithofacies from outcrop
- To delineate lithofacies succession from outcrop and sedimentologic log section
- To classify lithofacies into facies association and establish the Environment of Deposition (EOD).

1.5   Study Methodology

Ten outcrops (assigned numbers 1–10) from some recently and previously exposed rock units on road cut and mining pits (quarries) sections across Akpoha through Ngodo sandstone ridges of Afikpo were studied. The study involved mapping, logging, classification and sampling of outcropping lithologies of various lithostratigraphic units (Fig. 1.1). Information obtained from extent of exposed rock units and attitude of beds (strike, dip direction and dip amount) and rock type were used in producing a detailed geologic map of the study. Sedimentological logs were generated from data obtained from measured sections of exposed rock units and their field characteristics. Modified SedLog 2.1.4™ software was used in producing and interpreting the lithologic log motifs/profiles. Folk (1959, 1974) and Dunham (1962) systems were used in the classification of sandstones and carbonates respectively.

Collected rock samples were subjected to facies and petrographic analysis. Lithofacies analysis was carried out using key descriptive rock attributes, which include lithology (rock type), texture (grain size, grain shape and grain sorting), fossil content and sedimentary structure (erosional, syn-depositional, post-depositional and biogenic). Primary sedimentary structures were used to deduce the processes and conditions of deposition, the directions of the currents, which deposited the sediments, and in areas of folded rocks, the way-up of the strata. In addition, textural and mineralogical characteristics of lithofacies were interpreted from handheld samples and petrographic (thin section) analysis. Identified lithofacies within recognised facies succession (deduced from stacking pattern observed on log motifs) were grouped into facies association, which aided in the interpretation of environments of deposition. Paleo-depositional environment model was reconstructed from information obtained from measures outcrop sections and facies analysis. Generally, the workflow adopted for this study is shown in Fig. 1.4.

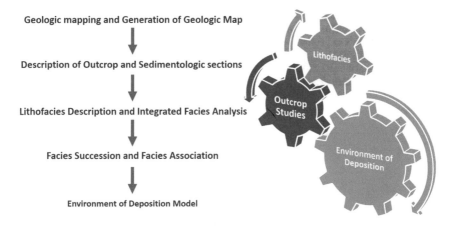

Fig. 1.4 Generated workflow for this research work

References

Amajor LC (1980) A subtidal bar model for the Eze-Aku sand bodies, Nigeria. J Sediment Geol 25: 291–309

Amajor LC (1987) The Eze-Aku Sandstone ridge (Turonian) of southeastern Nigeria: A re-interpretation of their depositional origin. J Min Geol 23: 17–26

Akpan EB, Nyong EE (1987) Trace fossils assemblages and depositional environment of Turonian calcareous sandstones in the Southeastern Benue Trough, Nigeria. J Afr Earth Sci 6(2):175–180

Banerjee I (1980) A subtidal bar model for the Eze-Aku Sandstones, Nigeria. J Sediment Geol 30:133–147

Dim CIP, Okwara IC, Mode AW, Onuoha KM (2016) Lithofacies and environments of deposition within the middle–upper cretaceous successions of southeastern Nigeria. Arab J Geosci 9:447. https://doi.org/10.1007/s12517-016-2486-7

Dim CIP, K. Onuoha KM, Okwara IC, Okonkwo IA, Ibemesi PO (2019) Facies analysis and depositional environment of the Campano – Maastrichtian coal-bearing Mamu Formation in the Anambra Basin, Nigeria. J Afr Earth Sci 152: 69–83

Dunham RJ (1962) Classification of carbonate rock according to depositional texture. In: Ham WE (ed) Classification of carbonate rocks: American Association of Petroleum Geologists Memoir, pp 108–121

Folk RL (1959) Practical petrographic classification of limestones. Am Assoc Pet Geol Bull 43:1–38

Folk RL (1974) Petrology of sedimentary rocks. Hemphill Publishing Company, Texas, Austin, p 182

Hoque M (1977) Petrographic differentiation of tectonically controlled Cretaceous sedimentary cycles, Southeastern Nigeria. Sediment Geol 17:235–245

Kendall CG (2005) Introduction to sedimentary facies, elements, hierarchy and architecture: a key to determining depositional setting. University of South Carolina. Spring, USC Sequence Stratigraphy Web Site. http://strata.geol.sc.edu

Lawal O (1991) Palynological age and correlation of a black shale section in the Eze-Aku Formation, Lower Benue Trough, Nigeria. J. Afr. Earth Sci 12:473–482

Miall AD (1996) The geology of fluvial deposits: sedimentary facies, basin analysis, and petroleum geology (first edition). Springer-Verlag, Berlin, p 582

Miall AD (2000) Principles of sedimentary basin analysis, 3rd edn. Springer-Verlag, Berlin Heidelberg, 616 p

Okoro AU, Igwe EO (2014) Lithofacies and depositional environment of the Amasiri Sandstone, Southern Benue Trough, Nigeria. J Afr Earth Sci 100:179–190

Simpsons A (1954) The geology of parts of Onitsha, Owerri, and the Nigerian coal fields. Geol Surv Niger Bull 24:121

Ukaegbu VU, Akpabio IO (2009) Geology and stratigraphy Northeast of Afikpo Basin, Lower Benue Trough, Nigeria. Pac J Sci Technol 10(1):518–527

Chapter 2
Regional Geology, Basin Forming Tectonics and Basin Fills of Southern Benue Trough and Anambra Basin in Afikpo Area

2.1 Geologic Framework

The cretaceous Benue Trough of Nigeria is the most interesting of the sedimentary basins in West Africa, chiefly because of the mid-Santonian folding movement that affected the marine and continental sediments within it. It contains up to 6000 m thick of alternating marine and fluvio-deltaic sediments ranging from Albian to Maastrichtian in age in places of its domain (Reyment 1965; Whiteman 1982; Wright et al. 1985). The works of Wright (1968), Burke et al. (1971), Grant (1971), Nwachukwu (1972), Olade (1975), Offodile (1976) and Benkhelil (1982) show that the Santonian inversion was associated with anticlinal (Abakaliki Anticlinorium) and synclinal (Afikpo Synclinorium) structures, and igneous emplacements. Hoque and Nwajide (1985) and Ofoegbu (1985) documented the sequence of events that led to the formation of the Trough namely; the rifting stage, trough stage, deformation stage (with magmatism) and the platform stage.

Previous studies reveal that the tectonic evolution of the Cretaceous Benue Trough was closely controlled by transcurrent faulting through an axial fault system, developing local compressional and tensional regimes and resulting in basins and basement horsts along releasing and restraining bends of the faults. The model of the tectonic evolution of the trough is based upon a general sinistral wrenching along the trough responsible for the structural arrangement and the geometry of the sub-basins Benkhelil (1989). Agagu and Adighije (1983) based on the tectonic and sedimentation framework suggested that the Aptian-Santonian southern Benue Trough was a relatively simple rift comprising a deep southern basinal area and a broad shallower platform to the north. Regional stratigraphy and sediment thickness patterns were thereafter strongly modified by the Lower Santonian tectonism in the Southern lower Benue Trough (Fig. 2.2).

© The Author(s), under exclusive license to Springer Nature Switzerland AG 2021
C. I. P. Dim et al., *Facies Analysis and Interpretation in Southeastern Nigeria's Inland Basins*, SpringerBriefs in Earth Sciences, https://doi.org/10.1007/978-3-030-68188-3_2

2.2 Regional Tectonic Setting

The Benue Trough is located almost exclusively in Nigeria except for the extreme tip of the Yola branch, which lies in Cameroon, is a NE-SW trending sedimentary basin about 1000 km long and 50–100 km wide that is situated at a major re-entrant in the West African continental margin (Whiteman 1982; Wright et al. 1985). The trough and extends from the Niger delta in the Gulf of Guinea to the Chad basin in the interior of the West African Pre-Cambrian shield (Fig. 2.1). The works of Burke et al. (1971), Nwachukwu (1972), Burke and Dewey (1973), Olade (1975, 1976) and Chukwu-Ike (1981) have extensively discussed the tectonic setting of the tectonic of the trough.

Geographical survey has revealed that wrenching was a dominant tectonic process during the Benue Trough evolution (Benkhelil 1982; Whiteman 1982). The model is based on the existence of an axial central positive gravity anomaly interpreted as a basement high, and is flanked by two linear negative anomalies. Although, several models had been proposed, the application of Y—shaped triple junction rift model (RRR) to the break-up of the Afro-Brazilian plate, during early Cretaceous times as documented by Burke (1976), Burke et al. (1970), Fitton (1983), Hoffman et al. (1974), Olade (1975, 1978, 1979), and Wright (1981) provides the best explanation of the Benue Trough configuration. It is envisaged that two arms of the RRR rift system each separated to form the South Atlantic, and the Benue Trough represents the third, the failed arm.

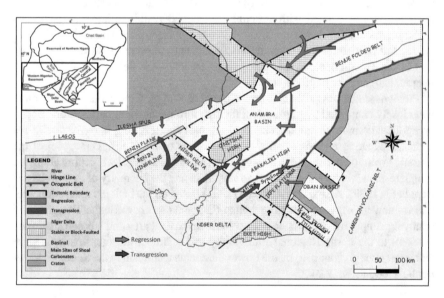

Fig. 2.1 Megatectonic and depositional framework of the southern Benue Trough after the Santonian showing the Abakaliki High/Anticlinorium, Anambra Basin and Afikpo Syncline (after Kogbe 1989)

The concept of origin of the Benue Trough led to its classification as an aulacogen (a linear, deeply subsiding trough extending from a continental margin or geosyncline into the interior of a foreland platform or craton) (Olade 1975). The works of Salop and Scheinmann (1969), Burke and Dewey (1973) and Burke and Whiteman (1973) led to the recognition of aulocogens throughout the world in both Proterozoic and Phanerozoic terrains (Wright 1968).

The basin fills of the Benue Trough (an intracratonic rift system) are thought to originate from partial relief of stress distortion, hence were wedged apart. Following the separation of Africa from South America plates in the Late Senonian (about 80 million years ago), the southern half of Africa tended to swing back, such that the stretched Benue Trough region was transformed into a minor compressional belt, folding the Pre-Santonian sedimentary fill of the trough (Wright 1968). Nwachukwu (1972) reported that early Cretaceous sediment packages, which were affected by the Turonian and Santonian events were more intensely deformed. Other views on the origin of the Benue Trough are found in the work of Wilson and Willians (1979) and Benkhelil (1982) that attributed its origin to the onshore extension of equatorial oceanic fracture zones along the northwestern and southeastern margins of the trough. Based on the study of igneous intrusives, pyroclastics and the stratigraphic sequence in southern Benue Trough, Hoque and Nwajide (1985) proposed a conceptual model of the evolution of the trough.

The Benue Trough is commonly divided into three main domains corresponding to both geological and geo-morphological partition.

(a) The Yola—Gombe Branch treading N55° E.
(b) The Middle Benue Trough, which is the linear part of the Basin.
(c) The Lower Benue Trough, which has shifted southwest, includes two main structures.

    (i) The N60° E trending Abakaliki Anticlinorium flanked by.
    (ii) The Anambra Basin and the Afikpo syncline (Fig. 2.3).

The geologic evolution of Southern Nigeria sedimentary basins was controlled by three major tectonic phases, which took place in the Albian, Santonian and the Late Eocene or early Oligocene times (Fig. 2.3). These major tectonic phases resulted to the displacement of the axis of the main basin giving rise to these three successive basins (Murat 1972): (a) The Southern Benue Trough, (b) The Anambra Basin and (c) The Niger-Delta (Fig. 2.3). However, in this work, the focus will be on the southern Benue Trough and the Anambra Basin, which underlies the study area (Fig. 2.2).

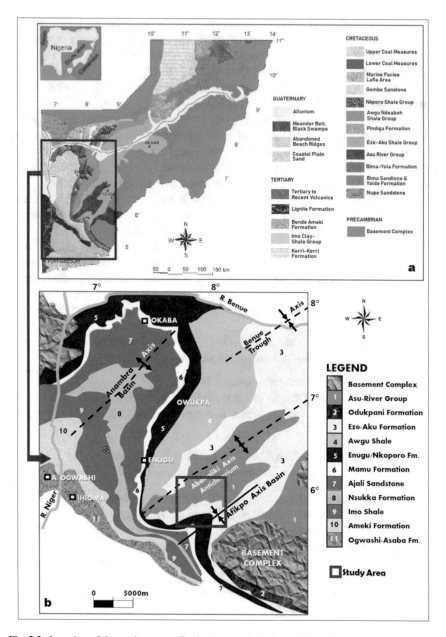

Fig. 2.2 Location of the study area. **a** Geologic map of the Benue Trough showing the Anambra Basin (after Uma 1998). **b** Geologic map of Anambra Basin showing the study area (in red box) (after Akande et al. 2007)

2.3 Regional Stratigraphic Setting

The stratigraphic successions of study area, which is situated within the southern Benue Trough and partly the lower intervals of the Anambra Basin have been documented in several works (Fig. 2.2; Petters 1978; Hoque 1976; Ojoh 1992). Although several literatures exist on the stratigraphy of the southeastern Nigerian basins, there still some varied thoughts from the generally accepted view that the Afikpo sub-basin and overlying Anambra Basin that is separated by an unconformity, are two separate entities. However, both are believed to have originated from the tectonic episode (Santonian inversion) that affected the trough since here are no noticeable structural demarcation between the two entities (Nwajide 2013). Hence, there are no justifiable basis upon which the Afikpo sub-basin should be accorded a basin distinct from the Anambra Basin. The northeastern part of the Benue Trough has a thick Cretaceous sedimentary rock succession, which is about 3500 m thick, covered in part by the Tertiary and Quaternary sediments of the Chad Basin (Allix and Popoff 1983). The thickness increases the southwestward to over seven thousand meters near the present Niger Delta (Hospers 1965), thus imparting a wedge-shaped geometry to the sedimentary body of the trough (Fig. 2.3).

The stratigraphic setting and the sedimentary fills of the southern (Lower) Benue, were controlled by cycles of transgressions and regressions and local tectonics. The depositional cycles were constant such that during the transgressions, shales, locally calcareous, were deposited in structural depression whereas coals and carbonates developed on submerged structural highs (platforms, horsts) protected from clastic influx (Murat 1970). Three such cycles of basin fill were recorded during the first tectonic phase (Fig. 2.4): (a) The Necomian-Cenomanian (Asu River Group), (b)

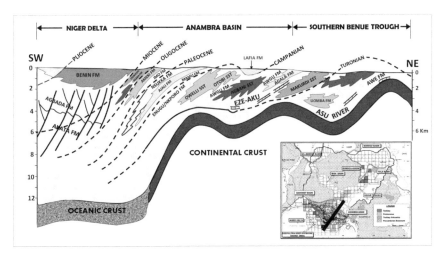

Fig. 2.3 Cross section of the showing lithostratigraphic units of Abakaliliki, Anambra and Niger Delta Basins (after Benkhelil 1986)

Generalized Stratigraphic Chart of the Southern Benue Trough

BASIN	FORMATION	AGE		ENVIRONMENT		DEPTH	SANDSTONE PETROLOGY	TECTONO-SEDIMENTOLOGIC STAGE
ANAMBRA	Nsukka Formation Ajali Formation	MAASTRICHTIAN		MARGINAL MARINE		0m	QUARTZ ARENITE	PLATFORM STAGE
	Mamu Formation Nkporo/Enugu formations	CAMPANIAN		SHELF		100m		
		SANTONIAN		FOLDING	MARGINAL MARINE			DEFORMATION STAGE
SOUTHERN BENUE TROUGH	AWGU GROUP (Awgu Formation/ Agbani sandstone/ Nkalagu Formation)	CONIANCIAN		MARINE				
		TURONIAN	UPPER	MARINE		1000m		
			MIDDLE	SHELF		1150m		
			LOWER	MARINE		1350m		
	EZE-AKU GROUP (Eze-Aku shale/ Agaila/Makurdi/ Amaseri sandstone/Ibir sandstone)	CENOMANIAN	UPPER	MARINE		1500m	FELDSPATHIC SANDSTONE	TROUGH STAGE
			MIDDLE	MIXED				
			LOWER	SUBCONTINENTAL		1880m		
	ASU-RIVER GROUP (Abakaliki shale/ Minor intrusions)	UPPER ALBIAN	LATE	NEARSHORE		1980m		
			MIDDLE	INTERNAL AND EXTERNAL SHELF				
			EARLY			2130m		
		MIDDLE ALBIAN		MARINE BASIN		3630m		
	NOT OUTCROPPING ?	PE-MIDDLE ALBIAN (Aptian, Neocomian)		DELTAIC				RIFTING STAGE
				NON MARINE		5000m		
	MAJOR DISCORDANCE			METAMORPHIC				
	PRECAMBRIAN BASEMENT							

After Ojoh, 1990; Petters, 1991 and Murat, 1970.	After Hoque and Nwajide, 1985

Fig. 2.4 Generalized stratigraphic chart of the southern Benue Trough and Anambra Basin (after Hoque and Nwajide 1985; Murat 1972; Ojoh 1992; Petters 1991; Dim et al. 2016)

The Early to Late Turonian (Eze-Aku Group) and (c) The Coniancian–Santonian (Awgu Group). However, the focus of the study is on the stratigraphic succession of the Mid–Upper Cretaceous sequence (Cenomanian to Campanian) of the southern Benue Trough and the overlying Anambra Basin outcropping in the Afikpo area (Figs. 2.4 and 2.5). A summary of the sandstone petrology and associated tectono-sedimentological evolution of the Southern Benue Trough and Anambra Basin lithostratigraphic succession are given in Fig. 2.4.

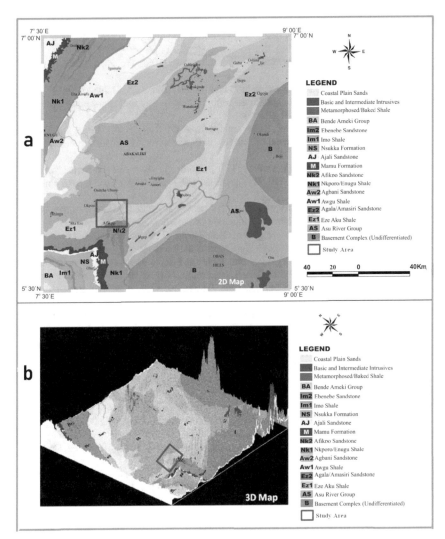

Fig. 2.5 2D and 3D geologic maps of the southern Benue Trough showing the lithostratigraphic units in the study area (adapted from Dim et al. 2014; Oha 2014)

References

Agagu OK, Adighije CI (1983) Tectonic and sedimentation framework of the lower Benue Trough, Southern Nigeria. J Afr Earth Sci 1:267–274

Akande SO, Ogunmoyero IB, Petersen HI, Nytoft HP (2007) Source rock evaluation of coals from the lower Maastrichtian Mamu Formation, SE Nigeria. J Pet Geol 30(4):303–324

Allix P, Popoff M (1983) Le crétacé inférieur de la partie nord orientale du fossé de la Bénoué (Nigeria) un exemple de relation étroite entre tectonique et sédimentation. Bulletin des Centres de Recherches Exploration-Production Elf-Aquitaine 7:349–359

Benkhelil J (1982) Benue Trough and the Benue Chain. Geol Mag 119:155–168

Benkhelil J (1986) Structure et evolution geodynamique du basin intracontinental de la Benoue (Nigeria), 3 These. University Nice, 226 p

Benkhelil J (1989) The origin and evolution of the Cretaceous Benue Trough, Nigeria. J Afr Earth Sci 8:251–282

Burke K (1976) The Chad Basin: an active intra-continental basin. Tectonophysics 36:197–206

Burke K, Dewey JF (1973) Plume-generated triple junctions: key indicators in applying plate tectonics to old rocks. Geology 81:406–433

Burke K, Whiteman AJ (1973) Uplift, rifting and the break-up of Africa. In: Tarling DH, Runcorn SK (eds) Implications of continental drift to the earth science, vol 2. Academic Press, London and New York, pp 734–755

Burke KC, Dessauvagie TFJ, Whiteman AJ (1970) Geological history of the Benue Valley and adjacent areas. In: Dessauvagie TFJ, Whiteman AJ (eds) African geology. Ibadan University Press, Ibadan, pp 187–205

Burke KC, Dessauvagie TFJ, Whiteman AJ (1971) Opening of the Gulf of Guinea and geological history of the Benue Depression and Niger Delta. Nat Phys Sci 233:51–55

Chukwu-Ike I (1981) Marginal fracture system of the Benue Trough in Nigeria and their tectonic implications. In: A Vogel (Ed-in-Chief) Earth evolutionary science, vol 2, pp 104–109

Dim CIP, Mode AW, Onuoha KM (2014) Lithofacies and depo-environmental implication: An outcrop study of Afikpo area in the southern Benue Trough, Nigeria. In: Unpublished Presentation Pack, 50th Annual International Conference and Exhibition, Nigerian Mining and Geosciences Society, Benin, 28 p

Dim CIP, Okwara IC, Mode AW, Onuoha KM (2016) Lithofacies and environments of deposition within the middle–upper cretaceous successions of southeastern Nigeria. Arab J Geosci 9:447. https://doi.org/10.1007/s12517-016-2486-7

Fitton JG (1983) Active versus passive continental rifting: evidence from the West African Rift system. Tectonophysics 94:473–481

Grant NK (1971) South Atlantic, Benue Trough and Gulf of Guinea Triple Junction. Geol Soc Am Bull 22:2295–2298

Hoffman P, Dewey JF, Burke K (1974) Aulacogens and their genetic relation to geosynclines, with a Proterozoic example from Great Slave Lake, Canada. Society of Economic Paleontologists and Mineralogists Special Publication, vol 19, pp 38–55

Hoque M (1976) Significance of textural and petrographic attributes of several Cretaceous sandstones, southern Nigeria. J Geol Soc India 17:514–521

Hoque M, Nwajide CS (1985) Tectono-sedimentological evolution of an enlongate intracratonic basin (aulacogen): the case of the Benue Trough of Nigeria. J Min Geol 21:19–26

Hospers J (1965) Gravity field and structure of the Niger Delta, Nigeria. Bull Geol Soc Am 76(4):407–422

Kogbe CA (1989) The Cretaceous Paleocene sediments of Southern Nigeria. In: CA Kogbe (ed) Geology of Nigeria. Rock View Ltd., Jos., pp 356–363

Murat RC (1970) Structural geology of the Cenozoic Niger Delta. In: African geology, vol 4. University of Ibadan Press, pp 635–646

Murat RC (1972) Stratigraphy and palaeogeography of the Cretaceous and Lower Tertiary in Southern Nigeria. In: Dessauvagie FJ, Whiteman AJ (eds) African geology. University of Ibadan Press, Ibadan, Nigeria, pp 251–266

Nwachukwu SO (1972) The tectonic evolution of the southern portion of the Benue Trough, Nigeria. Geol Mag 109:411–419

Nwajide CS (2013) Geology of Nigeria's sedimentary basins. CSS Bookshops Ltd, Lagos, p 565

Offodile ME (1976) The geology of the Middle Benue Trough, Nigeria. Bulletin of Geological Institution. University of Uppsala, Ph.D. Thesis, 166 p

Ofoegbu CO (1985) A review of the geology of the Benue Trough of Nigeria. J Afr Earth Sci 3:283–291

Oha IA (2014) Digital image processing and interpretation of Landsat 7 ETM+ data of part of the Lower Benue Trough, southeastern Nigeria. Unpublished Ph.D. Seminar Presentation Pack, University of Nigeria Nsukka, 37 p

Ojoh KA (1992) The southern part of the Benue Trough (Nigeria) Cretaceous stratigraphy, basin analysis, paleo-oceanography and geodynamic evolution in the equatorial domain of the South Atlantic. Nigerian Assoc Pet Explorationists' Bull 7(2):131–152

Olade MA (1975) Evolution of Nigerian Benue Trough (Aulacogen): A tectonic model. Geol Mag 112:575–583

Olade MA (1976) On the genesis of lead-zinc deposits in Nigeria's Benue rift (aulacogen): A re-interpretation. Niger J Min Geol 13:20–27

Olade MA (1978) Early Cretaceous Basaltic volcanism and initial continental rifting in Benue trough. Nat Phys Sci 27(3):458–459

Olade MA (1979) The Abakaliki pyroclastics of southern Benue Trough Nigeria: their petrology and tectonic significance. J Min Geol 16:17–26

Petters SW (1978) Mid-Cretaceous paleoenvironments and biostratigraphy of the Benue trough, Nigeria. Geol Soc Am Bull 89:151–154

Petters SW (1991) Regional geology of Africa. Springer Verlag, Berlin, p 722

Reyment RA (1965) Aspects of the geology of Nigeria: the stratigraphy of the Cretaceous and Cenozoic deposits. Ibadan University Press, Ibadan, p 145

Salop LI, Scheinmann YM (1969) Tectonic history and structures of platforms and shields. Teclnnophysics 7(5–6):565–597

Uma KO (1998) The Brine field of the Benue Trough, Nigeria a comparatively study of geomorphic, tectonic and hydrothermal properties. J Afr Earth Sci 26(2):261–275

Whiteman A (1982) Nigeria: its petroleum geology, resources and potentials, vol 2. Graham & Trotman Ltd., London, UK, 394 p

Wilson RCC, Willians CA (1979) Oceanic transform structures and the developments of Atlantic continental margin sedimentary basin a review. J Geol Soc Lond 136:311–320

Wright JB (1968) South Atlantic continental drift and the Benue Trough. Tectonophysics 6:301–310

Wright JB (1981) Review of the origin and evolution of the Benue Trough in Nigeria. Earth Evol Sci 2:98–103

Wright JB, Hastings DA, Jones WB, Williams HR (1985) Geology and mineral resources of West Africa. George Allen & Unwin Ltd., London, UK, 102 p

Chapter 3
Outcrop-Based Field Geologic Studies and Description of Measured Stratigraphic Successions

3.1 Field Geology/Description of Outcrop Location

Outcrop studies show that several sandstone ridges and valleys characterize the Afikpo area. The prominent sandstone ridges include those of Amaseri, Akpoha, Ibii, Ozara–Ukwu and Ngodo ridges (Fig. 1.1b, c). On the basis of orientation, the ridge belts were sub-divided into two, namely the NE–SW trending ridges paralleling the axis of the Benue Trough and the E–W ridges. Sand bodies occurring within the area are discontinuous (less than 1 km to few kilometers in length) and trend in E–Wand ENE–WSW directions. Quarrying activities on these sandstone ridges provided section for outcrop studies. Outcrop sections logged across the northern through central and southern part of the area summarized in Tables 3.1 and 3.2 and discussed below.

3.1.1 Amaseri Ridge Section (Outcrop 1)

This section is exposed on the Amaseri ridge through quarry activities (Crush Stone quarry site), along Amaseri–Okposi road (Fig. 1.1). The Amaseri section comprises stacked sediment thickness of 66.5 m. Lithologic units (predominantly sandstones and shales) extend from Amaseri through Oge Ubi to Asu River. Logged section revealed that the lower section of the outcrop is made up well-bedded sandstone units of varying thickness, while the intermediate shale unit and the upper sections comprise mainly sandstone units. A prominent feature observed in this section is an anticlinal plunged fold with tilted sandstone/shale beds, characterized by series of joint structures (Figs. 3.1 and 3.2).

© The Author(s), under exclusive license to Springer Nature Switzerland AG 2021
C. I. P. Dim et al., *Facies Analysis and Interpretation in Southeastern Nigeria's Inland Basins*, SpringerBriefs in Earth Sciences,
https://doi.org/10.1007/978-3-030-68188-3_3

Table 3.1 Summary of outcrop locations in the Amaseri through Ibii sections

Crush Stone Industry (Amaseri Ridge) quarry section	
Specifics	Latitude N 05° 56′ 00.4″; Longitude E 007° 53′ 12.3″
Regional/province/state/country	South East/Ezeke-Amaseri Area, Off Afikpo-Abakaliki/Okigwe-Abba Omega Axis/Ebonyi State, Nigeria
Formation name	Eze-Aku Formation (Amaseri Sandstone Member)
Age	Turonian (Pre-Santonian)
Thickness	67 m (219.76 ft.)
Lithology	Siltstone/Fine- to coarse-grained sandstone and Conglomerates; average medium-grained; poor to moderately sorting
Other	Anticlinal folding and joint structures associated with Santonian events
Julius Berger Nigeria, Plc. (Akpoha-Ridge) quarry section	
Specifics	Latitude N 05° 56′ 59.0″ and Longitude E 007° 56′ 24.2″
Regional/province/state/country	South East/Akpoha Area, along Afikpo-Abakaliki Road/Ebonyi State, Nigeria
Formation name	Eze-Aku (Amaseri Sandstone Member)
Age	Turonian (Pre-Santonian)
Thickness	70.6 m (231.568 ft.)
Lithology	Shale, Siltstone, Sandstone, Limestone
Other	Sedimentary structures present include, parallel lamination, intra-formational clast, minor folding, planar and trough cross-beddings
Along Akpoha-Ibi Ridge Quarry and Road-cut sections	
Specifics	Latitude N 05° 56′ 24.0″ and Longitude E 007° 55′ 10.0″
Regional/province/state/country	South-East/Akpoha-Ibi Area, along Afikpo-Abakaliki Road/Ebonyi State, Nigeria
Formation name	Eze-Aku (Amaseri Sandstone Member)
Age	Turonian (Pre-Santonian)
Thickness	71.5 m (234.52 ft.)
Lithology	Shale, Siltstone, Sandstone, Limestone
Other	Sedimentary structures present include, Load cast, biogenic structures

3.1.2   Akpoha Ridge Section (Outcrop 2)

This section is exposed on the Akpoha ridge through quarry activities (Julius Berger quarry site), Okigwe–Abba Omega road (Fig. 1.1). The Akpoha section comprises stacked sediment thickness of 70.3 m. Lithologic units (mainly sandstones, calcareous sandstone, shales, siltstones and limestones) extend from an artisan

Table 3.2 Summary of outcrop locations in the Ibii sections through Ngodo-Hill section

Marlum Nigeria Limited (Building and Civil Engineering)—Ibii Ridge quarry section

Specifics	Latitude N 05° 55′ 49.9″ and Longitude E 007° 55′ 25.9″
Regional/province/state/country	South East/Ibii Area, Off Afikpo-Abakaliki Road/Ebonyi State, Nigeria
Formation name	Eze-Aku (Amaseri Sandstone Member)
Age	Turonian (Pre-Santonian)
Thickness	110.5 m (362.44 ft.)
Lithology	Shale, Siltstone, Sandstone, Limestone
Other	Sedimentary structures present include, parallel lamination, intra-formational clast structures, biogenic structures

Ozara Ukwu section

Specifics	Latitude N 05° 53′ 59.1″ and Longitude E 007° 54′ 43.6″
Regional/province/state/country	South East/Opposite Mc gregor, along Afikpo-Abakiliki Road/Ebonyi State, Nigeria
Formation name	Eze-Aku (Amaseri Sandstone Member)
Age	Turonian (Pre-Santonian)
Thickness	221 m (724.88 ft.)
Lithology	Shale, Siltstone, Sandstone, Limestone
Other	Sedimentary structures include, planar, trough and, herringbone cross-beds, deformational band structures

Afikpo/Water-works section

Specifics	Latitude N 05° 53′ 44.2″ and Longitude E 007° 54′ 53.8″
Regional/province/state/country	South East/Opposite Mc gregor, along Afikpo-Abakiliki Road/Ebonyi State, Nigeria
Formation name	Eze-Aku (Amaseri Sandstone Member)
Age	Turonian (Pre-Santonian)
Thickness	36.9 m (121.032 ft.)
Lithology	Shale, Siltstone, Sandstone, Limestone
Other	Sedimentary structures include, dolerite intrusion (sill), channel-fill structure

Why-Worry spring/Road cut section

Specifics	Latitude N 05° 53′ 40.3″ and Longitude E 007° 55′ 15.0″
Regional/province/state/country	South East/Opposite Mc gregor, along Afikpo-Abakiliki Road/Ebonyi State, Nigeria
Formation name	Eze-Aku (Amaseri Sandstone Member)
Age	Turonian (Pre-Santonian)
Thickness	37.6 m (123.328 ft.)
Lithology	Shale, Siltstone, Sandstone, Limestone

(continued)

Table 3.2 (continued)

Other	Sedimentary structures present include, deformational bands, deformational bands
Ngodo Hill section	
Specifics	Latitude N 05° 53′ 51.9″ and Longitude E 007° 56′ 35.7″
Regional/province/state/country	South East/Ngodo Area near Afikpo Government College, off Afikpo-Abakaliki Road/Ebonyi State, Nigeria
Formation name	Eze-Aku (Amaseri Sandstone Member)
Age	Turonian (Pre-Santonian)
Thickness	53 m (173.84 ft.)
Lithology	Shale, Siltstone, Sandstone, Limestone
Other	Sedimentary structures present include, wave rippled surfaces, deformation structures and biogenic structures

local quarry along Okigwe–Abba Omega road through the Julius Berger quarry site along Abakaliki-Afikpo road. Logged section revealed that the lower section of the outcrop is made up of coarse-grained basal sandstone and shale units with some relatively thick, well-stratified calcareous sandstone/siltstone of varying thicknesses. The upper section comprises thick calcareous siltstone units with some thin heterolith (of siltstone/fine-grained and shale) units and thick medium to coarse-grained sandstone units. Well-developed joint structures characterized the exposed rock units of this section (Figs. 3.3 and 3.4).

3.1.3   Akpoha-Ibii Section (Outcrop 3)

This section is exposed through small-scale quarry activities found in-between the Akpoha and Ibii ridge sections along the Abakaliki-Afikpo Road. The Akpoha-Ibii section comprises stacked sediment thickness of 72 m. Lithologic units (mainly sandstones, bioturbated siltstone and shales) extend from Akpoha to Ibii. Logged section revealed that the lower section of the outcrop is made up of a very thick shale unit with relatively thin fine to medium-grained basal sandstone. The upper section comprise of a basal siltstone, capped by thick shale and sandstone interbeds (Fig. 3.5).

3.1.4   Ibii Ridge Section (Outcrops 4 and 5)

This section comprises of two outcrops, exposed on the Ibii ridge through quarry activities (Marlum Nigeria Limited quarry site and a local small-scale quarry site) along Abakaliki-Afikpo road (Fig. 1.1). The Ibii section comprises stacked sediment

Fig. 3.1 **a** A photomosaic showing a highly tilted units of the Amaseri ridge section (Eze-Aku Formation) in the western side of the studied area (Crush Stone Industry quarry site), North-west of Afikpo town off Amoso-Amaseri Rd. The dip for the sedimentary beds of the formation is approximately 25° NE? **b** High-angled tilted sediments with a prominent anticlinal fold structure exposed at Amaseri ridge section. **c** A close up on upper section, showing the limb of the fold (with well-bedded medium to coarse-grained sandstone units). **d** A close-up on the middle section showing the fold axis, with thick sediment package of well-stratified fine to coarse grained sandstone units. **e** A close up on the lower section showing sandstone package overlain by shale that is partly eroded. Tilted, laminated dark grey-brown silty-shale unit (person for scale approximately 1.8 m [5.9 ft] tall)

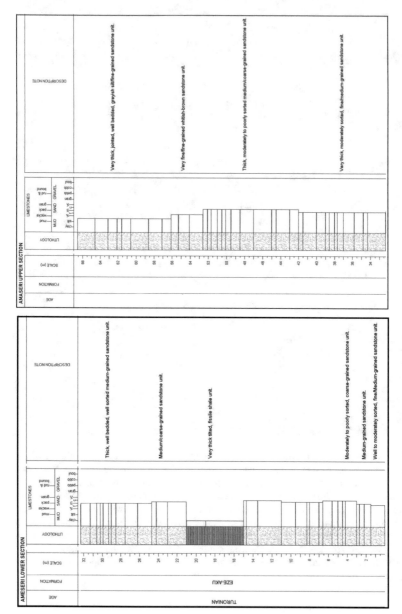

Fig. 3.2 Measured sedimentologic log of the lower and upper sections of Amaseri Ridge outcrop section

Fig. 3.3 a Thick well-bedded thick sandstone tilted section of highly fossilferous, calcareous silt to very fine-grained sandstone characterized by parallel laminated medium to coarse-grained sandstone, exposed at Akpoha Ridge section (Julius Berger Quarry site). **b** Sand and shale interstratification exposed at Akpoha Ridge section (Julius Berger Quarry site). **c** Joint section on siltstone unit and sandstone unit exposed at Akpoha Ridge section (Julius Berger Quarry site). **d** Well stratified tilted silty to fine-grained sandstone unit with highly fractured (joints) sections. **e** Joint contact between Coarse-grained Sandstone and fine-grained sand/Siltstone unit. *Note* **a–e** Outcrop exposure on in Akpoha sandstone ridge (Julius Berger quarry site). **f** are Quarry section about 100 m to Julius Berger Quarry along Abba Omega road, Afikpo North, southeastern, Nigeria. Person and geologic hammer for scale are approximately 1.8

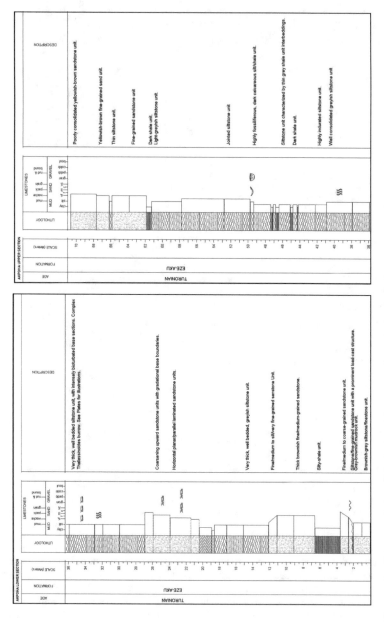

Fig. 3.4 Measured sedimentologic log of the lower and upper sections of Akpoha Ridge outcrop section

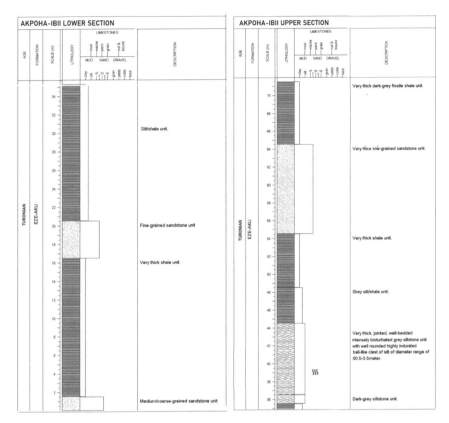

Fig. 3.5 Measured sedimentologic log of the lower and upper sections of Akpoha-Ibii outcrop section

thickness of 108.5 m. This thick sediment package conformably overlies the shale units of Akpoha–Ibii section. Lithologic units (mainly sandstones, calcareous sandstone, siltstones and shales) that are highly tilted. Logged section revealed that the lower section of the outcrop is made up of a very thick shale with thin interstratified fine to medium-grained sandstone units. The upper section comprises thick fine to medium-grained sandstone with relatively thick interstratified shale units (Figs. 3.6 and 3.7).

3.1.5 Ozara-Ukwu Ridge Sections (Outcrops 6 and 7)

This section comprises three sub-ridges/sections with exposed rock units separated by very thick shales. Outcrops 6 and 7 are exposed within these sub-ridges. Logged section across these three sections (Ozara-Ukwu I, II and III) shows sediment package thicknesses of 68, 72 and 56 m with an overall of 221 m thick. Lithologic units (mainly

Fig. 3.6 a Photomosaic of a tilted of a thick fine-grained sandstone to siltstone sediment at Ibii outcrop section (sandstone ridge) exposed at the eastern side of the study area (looking N-NE of Marlum quarry section). **b** Thick parallel laminated sandstone unit exposed at Marlum quarry section, Ibii. **c** A close-up on the eastern part of **a**, showing thick and well-stratified fine-grained sandstone units. **d** Large ball clast/boulder occurring as concretion on very thick fine-grained sandstone/siltstone units at Ibii outcrop section of Afikpo area, southeastern, Nigeria (after Dim et al. 2016). **e** A highly dipping, laminated and fissile shale (partly heterolithic) unit exposed at lower section of Ibii area (Dim et al. 2016). **f** Interstratification of sandstone and shale unit showing a thinning-upward sand and thickening-upward shale package

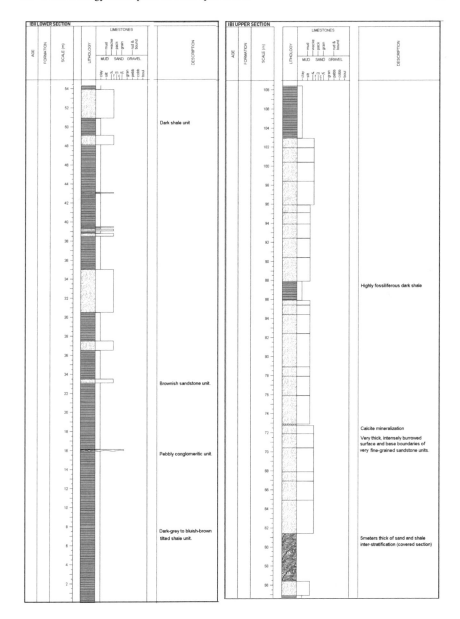

Fig. 3.7 Measured sedimentologic log of the lower and upper sections of Ibii outcrop section

sandstones and shales) extend from Ozara-Ukwu to Abakaliki-Afikpo/Okigwe–Abba Omega-Afikpo road junction. The outcropping units within these sub-sections are made of thick interstratified sandstone and shale units. The sandstone units are characterised by planar, trough and herringbone cross-beddings. Well-developed strike-slip fault structures and deformational bands on exposed rock units are prominent features seen in this section (Figs. 3.8 and 3.9).

3.1.6   *Ozara-Ukwu/Afikpo Section (Outcrop 8)*

This section is exposed on the boundary between Ozara-Ukwu and Afikpo area, along Okigwe-Afikpo road (Fig. 1.1). This section comprises stacked sediment thickness of 37 m. Lithologic units outcropping in this section includes conglomerates, sandstone, siltstones, shales, and dolerite. Logged section revealed that the lower section of the outcrop is made up of basal igneous intrusive (dolerite), which occurred as a sill (emplace concordantly with the sedimentary units). This is overlain by thick sandstone units with some interstratified shale units. A thick conglomeritic (pebbly) unit caps this section. In addition, fractured channel fill was observes at the upper part of this section (Figs. 3.10 and 3.11).

3.1.7   *Why Worry Spring–Afikpo Section (Outcrop 9)*

This section is exposed around Mcgregor College–Afikpo Junction (Fig. 1.1). The section comprises stacked sediment thickness of 37 m. Lithologic units outcropping in this section includes conglomerates, sandstones and shales. Logged section revealed that the lower section of the outcrop is made up of thick basal shale unit overlain by coarse-grained to conglomeritic sands at the upper section (Fig. 3.12).

3.1.8   *Ngodo Hill Section (Outcrop 10)*

This section is exposed on is exposed at Ngodo town, few kilometers from McGregor College–Afikpo Junction (Fig. 1.1). The section comprises stacked sediment thickness of 53 m. Lithologic units outcropping in this section includes conglomerates, sandstones and shales. Logged section revealed that the lower section of the outcrop is made up of thick basal shale unit overlain by coarse-grained to conglomeritic sands at the upper section (Figs. 3.13 and 3.14).

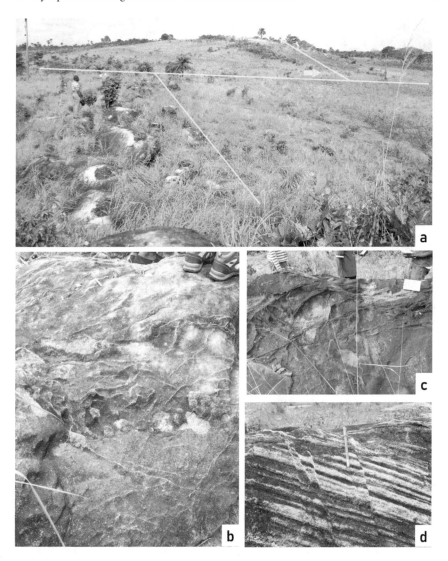

Fig. 3.8 **a** Well-developed strike-slip fault on the Ozara-Ukwu sandstone ridge, along Okigwe-Afikpo Road (*Note* Geologist standing on the one arm of the displace block). **b** Anastomosing deformational bands on sandstone unit. **c** Ferrogenized anastomosing deformational bands on rock units. **d** Striation marks observed on fault shear zone, along Okigwe-Afikpo Road

3.2 Synopsis of Lithologic Units and Their Field Characteristics

The influence of structural deformation (due to the Santonian folding) is evident on the stratigraphic succession as beds are tilted. The distribution of outcrop locations

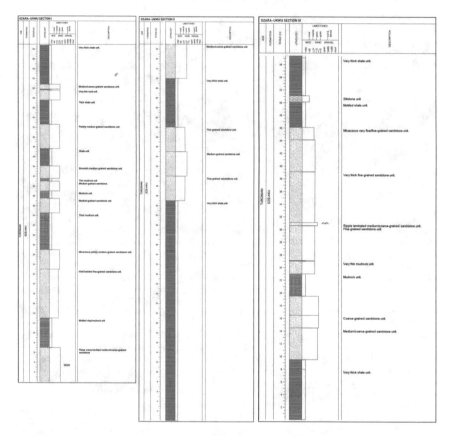

Fig. 3.9 Measured sedimentologic log of the Ozara-Ukwu outcrop sections

is such that the measured tilted stratigraphic succession overlies each other when stacked together from the northern to the southern part of the study area (Fig. 3.15a, b). Measured attitude of observed beds indicates that stratigraphic packages trend east to west (54° ENE–234° WSW) and dip southerly (144° SSE) with a variable range of dip amounts of 25°–40° in the northern and 9°–7° in the southern sections. Outcropping lithologic units/rock type that predominates across the study area includes; shale, siltstone, sandstone (some are calcareous), conglomerate, limestone and dolerite. Occasional sub-oval clast rip-out and insitu clasts with some in-place boulders are seen on the bedding plane of sandstone units (Figs. 3.6d and 3.15b). The thickness of beds of these rock types varies from as low as 0.25 m to as high as 2.3 m. Visible in the area is the high frequency occurrence of shale and sandstone interstratification.

Primary sedimentary structures (parallel and wave-ripple lamination and tabular, trough, herringbone and hummocky cross-beddings) and secondary sedimentary structures (convolute lamination and load cast) were seen occurring on sandstone units (Figs. 3.13f and 3.15c, d). Structural features that are evident in the area are anticlinal fold plunge observed at the Amaseri section and fold limbs associated with

Fig. 3.10 **a** Channel-fill structure showing a well-laminated coarse-medium grained sandstone with conglomeritic sand at the upper section exposed along Afikpo area. **b** Highly tilted sandstone unit with thin shale interstratification exposed through road-cut along Abakaliki-Afikpo road. **c** Sharp boundary context between highly tilted sandstone and shale unit exposed through road-cut along Abakaliki-Afikpo road. **d** Highly dipping sandstone units exposed on Ozara-Ukwu ridge along Abakaliki-Afikpo road. **e** Dolerite sill at Afikpo/Water-works outcrop section showing conformable contact with the overlying shale unit

joint structures and deformational bands seen at Ozara-Ukwu sections, and igneous emplacement at Afikpo/Waterworks section (Figs. 3.1a, 3.8b–d and 3.15e, g). These joints and fault structures, and the igneous emplacement (dolerite sill) that is concordant with the host rock (shale) are secondary/post-depositional. Biogenic structures were also observed as some of the sandstones were intensely and partly bioturbated, and the shales are partly fossiliferous (Fig. 3.15h).

Based on the generated geologic map (Fig. 1.1), outcrops 1–5 are exposed at the northern within the Amaseri, Akpoha and Ibii ridges, where the sediment packages of

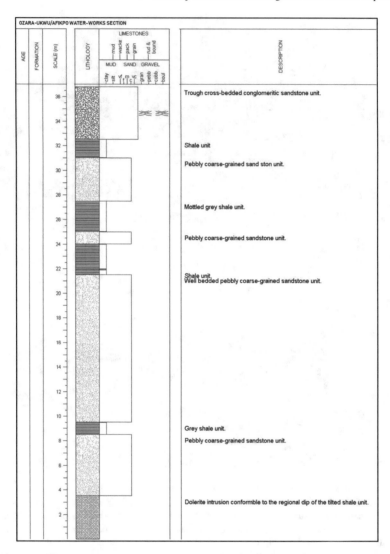

Fig. 3.11 Measured sedimentologic log of the Ozara-Ukwu/Afikpo outcrop section

Eze-Aku (Eze-Aku Shale and Amaseri Sandstone) lithostratigraphic unit (Southern Benue Trough) are the underlying rock units. Outcrop 6 and 7 are exposed at the central part with in the Ozara-Ukwu ridges where also, the sediment packages of Eze-Aku (Eze-Aku Shale and Amaseri Sandstone) lithostratigraphic unit (Southern Benue Trough) are the underlying rock units. However, outcrops 8–10 are exposed at the southern part within the sandstone ridges of Ngodo Hills where the sediment packages of Afikpo Formation (Anambra Basin) are the underlying rock units.

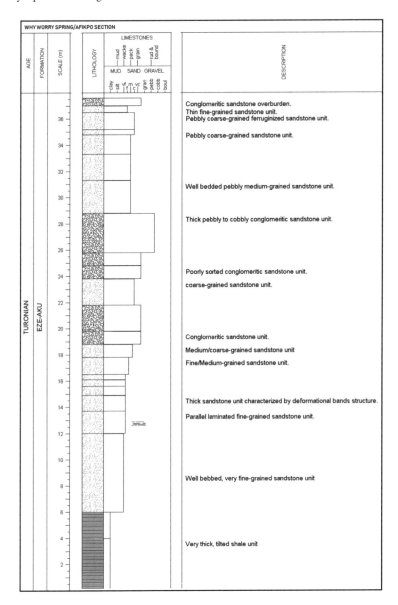

Fig. 3.12 Measured sedimentologic log of the Afikpo/Why Worry outcrop section

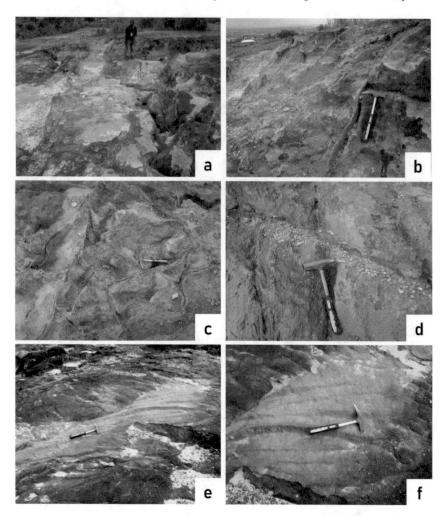

Fig. 3.13 **a** Ngodo Hill section with several anastomosing fractures on muddy conglomeritic unit. **b** Exposed section on Ngodo Hill showing joint structures on pebbly sandstone unit. **c** Anastomosing joint structures on hand muddy conglomeritic unit at Ngodo Hill. **d** Conglomeritic lag observed within sediment package at Ngodo Hill, Afikpo. **e** Wave rippled structure on sandstone unit. **f** A close-up on the series of rippled surface observed on pebbly sandstone unit

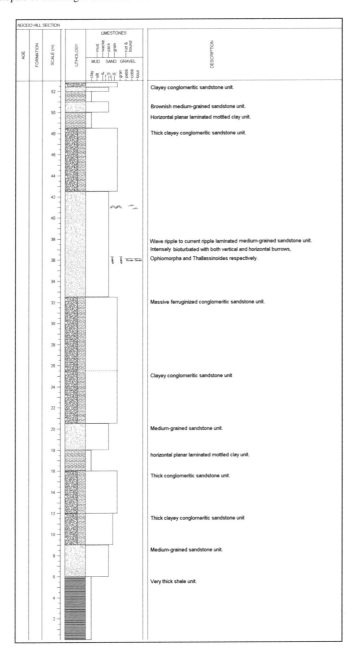

Fig. 3.14 Measured sedimentologic log of the Ngodo outcrop section

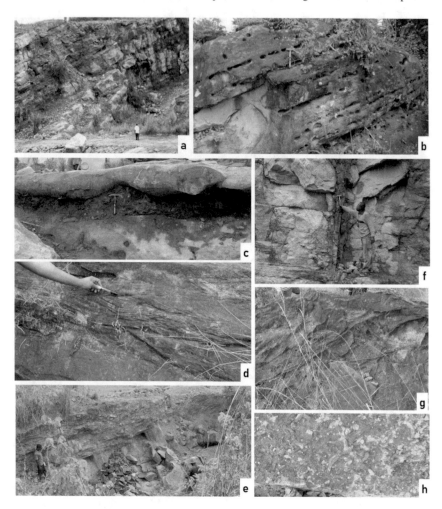

Fig. 3.15 a Dipping fold limb, with thick sediment package of well-stratified fine to coarse grained sandstone unit at Amaseri section. **b** Sandstone unit characterized by vugs (created by silt/mud clast) parallel to bedding plane at Akpoha-Ibii section. **c** Load structure (soft sediment deformation) at basal part of Akpoha outcrop section. **d** Trough cross-bedding structure on sandstone unit at Ozara-Ukwu section. **e** Dolerite sill at Afikpo/Water-works outcrop section. **f** Joint structure with infilling of indurates rock on a dipping silty-sandstone unit at Akpoha outcrop section. **g** Deformational bands on pebbly/conglomeritic sandstone unit exposed at Afikpo outcrop section. **h** Complex network of horizontal burrows—*Planolites* (*Cruziana* ichnofacies)

Reference

Dim CIP, Okwara IC, Mode AW, Onuoha KM (2016) Lithofacies and environments of deposition within the middle–upper cretaceous successions of southeastern Nigeria. Arab J Geosci 9:447. https://doi.org/10.1007/s12517-016-2486-7

Chapter 4
Analysis and Interpretation of Outcropping Facies

4.1 Facies Analysis

4.1.1 Lithofacies and Facies Succession

Detailed analyses of stratigraphic succession within the study area unravelled several genetically related facies that depict differences in the conditions (physical, biologic and chemical) that shaped their environments during the time of deposition. In this study, ten lithofacies deposited across continental, proximal marine and distal marine strata of the Eze-Aku Formation and Afikpo Formation outcropping on the sandstone ridges of Afikpo area were recognized, and their depositional enviroments were interpreted (Table 4.1). These facies and their subfacies as discussed below include medium–coarse and fine sandstone facies that are presented first, followed by limestone, mudstone and bioturbated siltstone and sandstone facies, and then the parallel and rippled laminated sandstones with the cross-bedded sandstone facies and finally the conglomeritic sandstone facies (Figs. 4.1, 4.2, 4.3, 4.4, 4.5, 4.6 and 4.7). These lithofacies are characterised by variable thicknesses that range from few centimeters to meters.

Log motifs/profiles from measured stratigraphic sections revealed that these lithofacies were deposited within two major facies cycles/successions, namely the shallowing/shoaling-upward cycle or coarsening/thickening-upward succession and the deepening upward cycle or fining/thinning-upward succession. Basal shale and mudstone facies or finer grained deposits overlain by sandstone facies or coarser grained deposits characterize the shoaling-upward cycle whereas basal sandstone facies or coarser grained deposits overlain by shale and mudstone facies or finer grained deposits characterize the deepening upward cycle. Furthermore, analysis of these log motifs shows the predominance of deepening upward cycle among the stratigraphic sections of the northern and central part of the study area, contrary to the southern part, which shows the dominance of shoaling-upward cycle (Fig. 4.1).

© The Author(s), under exclusive license to Springer Nature Switzerland AG 2021
C. I. P. Dim et al., *Facies Analysis and Interpretation in Southeastern Nigeria's Inland Basins*, SpringerBriefs in Earth Sciences,
https://doi.org/10.1007/978-3-030-68188-3_4

Table 4.1 Lithofacies occurrence and characteristics, facies association and depositional process interpretation in the Afikpo area and environs, southeastern, Nigeria (after Dim et al. 2016)

	Lithofacies and subordinate lithofacies	Facies association	Characteristics	Occurrence	Depositional processes and interpretation
1	Medium—coarse sandstone (MCS)	FA 2	Medium to coarse-grained sandstone, well stratified and partly massive. These are associated with sandstone units and shows convolute laminated structure, which is an evidence of soft-sediment deformation	Abundant	Deposited rapidly by such events as grain flow and turbidity currents and storms events (Miall 2000; Nichols 2009)
2	Fine sandstone (FS)	FA 3	Well sorted very fine to fine-grained calcareous sandstone units. These are associated with siltstone and shale units	Common	Rapid deposition from sediment-laden flows (Nichols 2009)
3	Bioclastic Limestone (BcL)	FA 4	Grey to dark grey fossiliferous limestones with substantial amount of calcareous shell fragment. BcL comprises of sandy skeletal grain-rich packstone with subequal spar and micrite content	Partly minor	Patch reef deposits washed by wave and tide action. Lagoonal to near shallow (Proximal) marine shelf settings (Spring and Hansen 1998; Boggs 2001)

(continued)

Table 4.1 (continued)

	Lithofacies and subordinate lithofacies	Facies association	Characteristics	Occurrence	Depositional processes and interpretation
4	Mudstone (MS)—(Silty-shale and Shale sub-lithofacies)	FA 1	Dark-grey silty-shale with thin siltstone to very fine-grained sandstone interlamination; Laminated clay shale with millimeter-scale laminae	Abundant	Deposits mostly within oxygenated, low energy setting, probably below fair-weather and storm wave base (Boggs 2001), in marine setting (Reyment 1965; Banerjee 1980; Pemberton et al. 1992)
5	Bioturbated Siltstone and sandstone (BSt)	FA 2	Bioturbated silty shale, siltstone, and very-fine to fine-grained sandstone. Intensity of bioturbation increases near bed tops; beds are completely burrowed by *Planolites* ichnofossils (*Cruziana* Ichnofacies)	Common	Deposits below fairweather wave base in marine setting. *Cruziana* ichnofacies, reflecting the work of an equilibrium community developed under fully marine conditions (Frey et al. 1990; Pemberton et al. 1992; MacEachern and Pemberton 1994)
6	Bioturbated Sandstone and sandstone (BSs)	FA 2	Bioturbated fine to coarse-grained sandstone, characterized by the presence of *Thallassinoides* and *Ophiomorpha* burrows that belongs to the *Glossifungites and Skolithos* ichnofacies respectively	Minor	Sediments deposited below fair weather wave base (Frey et al. 1990; Pemberton et al. 1992)
7	Planer-parallel laminated sandstone (PPLS)	FA 3	Light grey to brownish very fine to coarse-grained sandstone, with thin bedded planer/parallel laminated sandstones	Common	Planer bed flow, deposition under upper flow-regime, plane-bed conditions in marine setting (Miall 1978)

(continued)

Table 4.1 (continued)

	Lithofacies and subordinate lithofacies	Facies association	Characteristics	Occurrence	Depositional processes and interpretation
8	Cross-bedded sandstone (CBS)—(Tabular/Planar; Trough; Herringbone and Hummocky cross-bedded sandstone sub-lithofacies)	FA 4	Medium to very coarse-grained sandstone and partly pebbly. CBS lithofacies are characterized by solitary to grouped planar, trough, herringbone and hummock cross-beds in sets a few decimeters to 1 m thick	Partly common	Dunes (Lower Flow Regime), (Miall 1978, 1996). Bars deposits within tidal inlets
9	Ripple laminated sandstone (RLS)	FA 5	Very fine to very coarse and partly pebbly. These are characterized by ripple cross laminated structures	Minor	Lower Flow Regime (Miall 1978). Oscillatory wave processes that dominate in the upper shoreface environment, above fair-weather wave base (Gowland 1996)
10	Conglomeritic Sandstone (CS)	FA 5	Muddy conglomeritic sandstone associated with coarse-grained sandstone containing dispersed pebbles. Characterized by abundant *Thalassinoides* (*Glossifungites* ichnofacies)—bearing muddy sandstone and *Ophiomorpha* (*Skolithos* ichnofacies)—bearing sandstone	Minor	Transgressive lag deposits above firmground in a shallow sandy shoreline (Pemberton et al. 1992; MacEachern and Pemberton 1994)

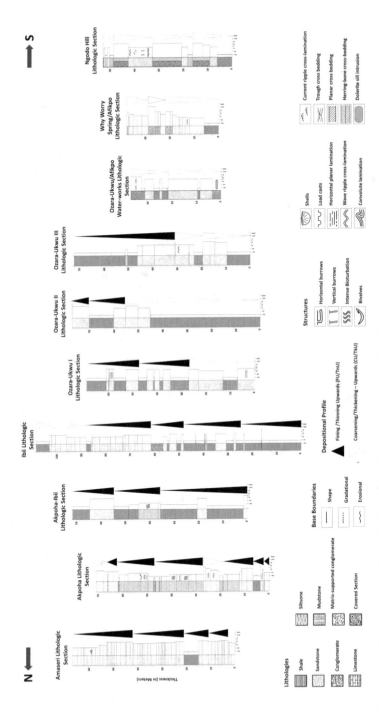

Fig. 4.1 Detailed measured sedimentological logs and stacking pattern/stratigraphic succession across Amaseri – Akpoha – Ibii – Ozara-Ukwu – Afikpo – Ngodo Hill outcrop sections (N–S), in Afikpo area and environs, southeastern Nigeria (Dim et al. 2016). *Note* Thickness is in meters

Fig. 4.2 a Thick sediment package of well-stratified medium to coarse-grained sandstone unit (MCS—Facies), exposed at Crust Stone Industry quarry section, Amaseri. **b** Convolute structure associated with soft sediment deformation exposed at Amaseri section. **c** Well-stratified fine-grained sandstone unit (FS Facies) exposed at Marlum Nigeria Limited quarry section Ibii. **d** Fine-grained sandstone unit (FS Facies), exposed at Crust Stone Industry quarry section, Amaseri

The medium–coarse sandstone facies is abundant in outcrops of Amaseri, Akpoha and Ibii where they occur at the basal and upper intervals of the stratigraphic sections. This facies is dominantly grey to brownish fine, medium and coarse-grained sandstone with bed thickness that varies from 0.1 to 0.5 m. There are little or no internal structures, although, minor plane-parallel lamination and convolute lamination are

Fig. 4.3 **a** Limestone bed (BcL Facies) associated with siltstone unit exposed at Akpoha section. **b** Limestone bed characterized by substantial amount of calcareous shells fragment of the bivalves and brachiopods. **c** Bioclastic limestone unit (with shells of bivalves—BsL Facies). **d** Close-up on hand-held broken piece of bivalve mold fro the limestone bed. **e** Calcite mineralised fracture associated with limestone units, exposed at Akpoha section

Fig. 4.4   a Thick muddy siltstone (MS Facies) exposed at Ibii Section. **b** Tilted, laminated dark grey-brown silty-shale unit (MS Facies), exposed at Marlum Nigeria Limited quarry site, Ibii. **c** A highly dipping, laminated and fissile shale unit exposed at lower section of Ibii area

seen in the fine-grained component of this facies (Fig. 4.2a, b). This facies are characterized by sharp basal contact and are seen within deepening upward cycles or fining/thinning upward facies succession. Observation from thin section shows subrounded mineral grains with high percentage of quartz mineral with relatively low feldspar content (Fig. 4.8a–e). Miall (2000) and Nichols (2009) suggested that the presence of this soft-sediment deformation structure (convolute lamination) in the medium–coarse sandstone facies is an indication of deposits associated with turbidity currents or storm.

The fine sandstone facies is common and is well seen at the outcrops of Ibii stratigraphic sections. This facies consists dominantly of very fine to fine-grained sandstone associated with siltstone and dark grey to bluish-grey shale units (Figs. 4.1 and 4.2c, d). The sandstone bed thickness ranges from 0.2 to 2 m thick. Interlaminated fine sands and shales are minor component of this facies. Sandstone are characterised by parallel laminations with some concentric lamina observed on boulder surfaces. This sequence of sedimentary structure observed in the fine sandstone are interpreted as probably rapid deposition from sediment-laden flow (Walker 1985; Nichols 2009).

Bioclastic limestone facies is minor in outcrop occurring only within the stratigraphic section of Ibii and Akpoha. This facies is typically grey fissiliferous, mud-rich limestone bed of 0.1–1 m thick (Figs. 4.1 and 4.3a–e). Silty sandstone units

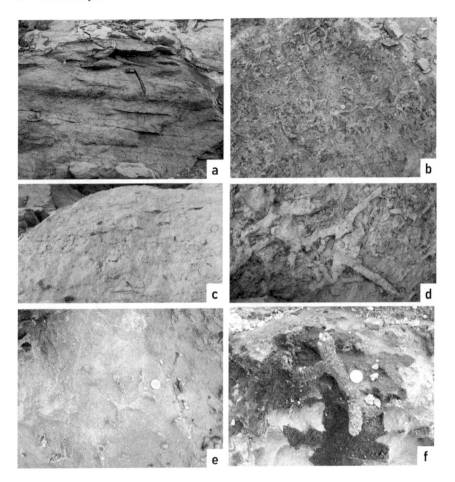

Fig. 4.5 **a** Siltsone with Highly bioturbated section (BSt Facies) exposed at Ibii section. **b** Complex network of horizontal burrows—*Planolites* (*Cruziana* ichnofacies), (BSt Facies), exposed at Ibii section. **c** Ophiomorpha burrows on silty-sandstone units exposed ta Akpoha section. **d** A close-up on complex network of horizontal burrows—*Planolites* (*Cruziana* ichnofacies), (BSt Facies), exposed at Ibbi section. **e** Horizontal burrows of *Thallassinoides* on conglomeritic sandstone (CS Facies), exposed at Ngodo Hill section **f** Vertical burrows *Ophiomorpha* on pebbly medium-grained sandstone unit (CS Facies), exposed at Ngo Hill section

are minor component of this facies. Field observation and thin section interpretation shows that bioclastic limestone facies is grain-supported and contain molds and calcite-mineralized shells of bivalves and brachiopods (Fig. 4.8h). Based on Dunham (1962) scheme these limestones where classified and interpreted as sandy skeletal grain-rich packstone allochems with sub-equal spar and micrite content. Boggs (2001) suggests that this facies were probably deposits within the shallow marine shelf environment.

Fig. 4.6 a Parallel laminated fine to medium-grained sandstone unit (MCS and PPLS Facies), exposed at Julius Berger Quarry site Akpoha. **b** Planar/parallel laminated sandstone unit (PPLS Facies), exposed at Ozara-Ukwu section. **c** Wave rippled structure on sandstone unit (RLS Facies), exposed at Ngodo Hill section

Mudstone facies is abundant in outcrop. The mudstone facies was classified into two sub-facies namely the silty shale and the shale facies seen at the basal parts of Amaseri, Akpoha, Ibii and Ozara–Ukwu outcrops (Figs. 4.1 and 4.4a–c). The silty shale subfacies is characterized by dark grey shale with siltstones intercalations. This subfacies is also referred to as the heterolith sub-facies due to the shale and siltstone intercalations. Fresh surface shows mottled appearance. In some outcrop sections, this sub-facies shows some well-developed shale and siltstone interlamination. The shale subfacies is characterised by dark-grey to brownish fossiliferous and calcareous shales with partly preserved faintly visible lamination. This subfacies are seen in the basal and upper parts of Ozara–Ukwu, Afikpo and Ngodo Hill outcrops. Generally, the thickness of mudstone facies ranges from 0.5 to 4 m. The mudstone facies are interpreted to be deposits from suspension in low-energy setting possibly below the storm wavebase and the mottled appearance is an indication of well-oxygenated setting (Reyment 1965; Banerjee 1980; Boggs 2001).

Bioturbated siltstone facies is common in outcrop. Bioturbated siltstone facies is dominantly bioturbated siltstone and very fine-grained sandstone. Siltstone of this facies is grey and intensely bioturbated. Associated light grey to brown calcareous sandstone ranges from very fine-grained to fine-grained with parallel lamination.

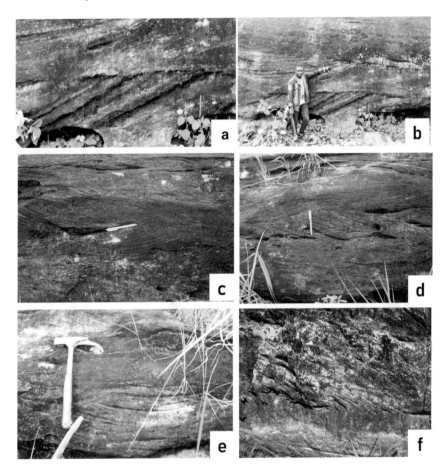

Fig. 4.7 **a** Tablar/planar cross-bed (CBS sub-facies), exposed at Government Technical College, Akpoha section. **b** Planar and Trough cross-beds (CBS sub-facies) on sandstone unit exposed at Government Technical College, Akpoha section. **c** Herringbone cross-bedding structure on sandstone unit (CBS sub-facies). **d** Herringbone cross-bedding structure on sandstone unit (CBS sub-facies), exposed at Ozara-Ukwu section. **e** Hummocky cross-bedding structure on sandstone unit (CBS sub-facies), exposed at Ozara-Ukwu section. **f** Planar/parallel laminated sandstone unit (PPLS Facies), exposed at Afikpo/Water-works section

Sandstone and siltstone bed thickness varies from 0.1 to 0.5 m. *Skolithos* and *Plano-lites* fossil traces and linear grove belonging to *Skolithos* and *Cruziana* association, are observed on siltstone and sandstone of this facies seen at the basal and upper parts of Akpoha and Ibii section (Figs. 4.1 and 4.5a–d). These ichnofossils are interpreted to be softground substrate-controlled, found in high and medium energy marine setting (Pemberton et al. 1992). Evidence from thin section shows the occurrence of relatively high percentage of feldspar minerals in the sandstone units of this facies (Fig. 4.8i).

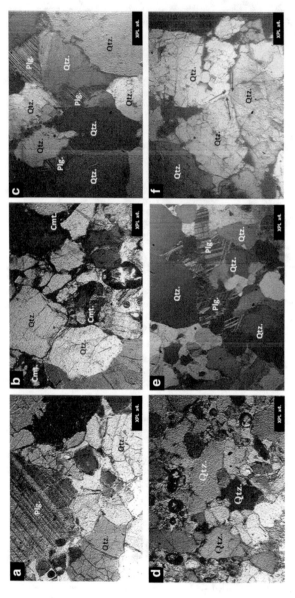

Fig. 4.8 **a** Photomicrograph of a pebbly poorly sorted, medium to coarse—grained sandstone, plagioclase (well-developed polysynthetic twinning) and quartz grain show fractures due to deformation. **b** Photomicrograph of a moderately sorted, medium to coarse—grained sandstone with quartz grain and accessory mineral forming cements. **c** Photomicrograph of a moderately sorted, coarse—grained sandstone with angular to sub angular minerals. **d** Photomicrograph of a sub-rounded, moderately sorted, medium-grained sandstone. **e** Photomicrograph of a moderately sorted, fine to medium-grained sandstone. **f** Photomicrograph of a well sorted, quartz-rich, coarse—grained, sandstone showing the presence of sutured grain contact and quartz grain, which displays dark streaks and healed fractures due to deformation. **g** Photomicrograph of a poorly rounded, well sorted, fine-grained sandstone with feldspars minerals; **h** Photomicrograph of a bioclastic limestone with calcite-filled bivalve shell that shows floating grain contact. **i** Photomicrograph of moderately sorted, fine—grained siltstone showing some partial crystal alignment (*NB*: **a–d** photomicrograph = Thin section of rock samples Amaseri outcrop section; **e–g** photomicrograph = Thin section of rock samples from Akpoha outcrop section; **h–i** photomicrograph = Thin section of rock samples from Ibii outcrop section. Abbreviations: Qtz. = quartz, Plg. = plagioclase and Mrc = microcline, Bt. = Biotite, Cmt. = Cements, Sh. = shell. Thin sections are displayed in XPL = Cross-Polarized Light) (after Dim et al. 2016)

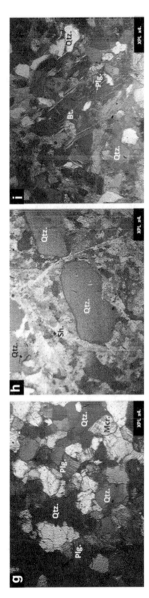

Fig. 4.8 (continued)

Bioturbated sandstone facies is common in the southern part but rare in the northern part of the area. This facies was seen at the outcrops of Afikpo and Ngodo Hill (Figs. 4.1 and 4.5d–f). Bioturbated sandstone facies consists of medium to coarse-grained sandstone occurring in beds of up to 1.0 m thick that are characterized by *Thallassinoides* and *Ophiomorpha* burrows. The sandstone of this facies includes beds of yellowish to reddish brown coarse-grained to conglomeritic sandstone. These *Thallassinoides* and *Ophiomorpha* burrows were limited to the upper few millimeter of sandstone units and the degree of bioturbation ranges from very slight to moderate. These ichnofossils of *Glossifungites* and *Skolithos* association suggest that they are dwelling burrows of deposit and suspension feeder found within the shoreface to offshore marine setting, and the bioturbated sandstone facies depicts sediments deposited below fair weather wave base (Pemberton et al. 1992; MacEachern and Pemberton 1992; LePain et al. 2009).

Plane-parallel laminated sandstone facies is common, and was seen at the outcrops of Akpoha, Ibii and Ozara–Ukwu. This facies consists of grey to brown fine and medium-grained sandstone in thin, planar, parallel beds from 0.01 to 0.20 m thick. Sandstone units of this facies are commonly internally laminated (laminae 0.02–0.15 thick), with little or no bioturbation (Fig. 4.4a–b). This facies characterize deposits of the shoaling upward cycle (Fig. 4.1). Harms (1979), Harms et al. (1982), Miall (1978, 1996) and Blatt et al. (1980) suggested that the plane-parallel laminated sandstone facies are deposits of upper flow-regime, typical of plane-bed conditions in marine environment.

Rippled laminated sandstone facies was quite rare occurring only within the outcrops of Ozara–Ukwu and Ngodo Hill. This facies is characterized by ripple laminated fine-grained sandstones with some clayey pebbly to conglomeritic intervals (Figs. 4.1 and 4.6c). Sandstone bed thickness of this facies varies from 0.1 to 3.5 m. There were no evidence of bioturbation. Miall (1978, 1996) and Blatt et al. (1980) suggested that the rippled laminated sandstone facies reflects deposits formed at lower flow regime in a marine setting.

Cross-bedded sandstone facies is a minor component, although appears to be common in the outcrops of Akpoha, Ozara-Ukwu and Afikpo stratigraphic section. This facies consist of grey to brown, fine to coarse-grained sandstones. Generally, the bed set thickness of the cross-bedded sandstone facies varies from 0.15 to 2.0 m. The cross-bedded sandstone facies was classified into four sub-facies namely; planar or tabular, trough, herringbone and hummocky cross-bedded sandstone facies (Fig. 4.8a–c, f).

(i)   *Tabular cross-bedded sandstone sub-facies* is characterized by planar–tangential foresets in single sets or multiple sets. These sub-facies result from the migration of large two-dimensional bedforms (straight to slightly sinuous crestlines) under upper lower flow-regime conditions (Harms et al. 1975; Collinson and Thompson 1989; Boggs 2006).

(ii) *Trough cross-bedded sandstone sub-facies* consist of fine to medium-grained trough cross-stratified sandstone that are characterized by tangential foresets.

The thickness of bed sets of cross-strata varies from 0.2 to 1 m. This sub-facies is typical of sediment deposited during the migration of bedforms that developed in a subaqueous setting under upper lower flow-regime conditions (Harms 1979; Harms et al. 1982; LePain et al. 2009).

(iii) *Herringbone cross-bedded sandstone sub-facies* are characterised by alternating cross-lamina forests dipping in opposite directions. This sub-facies suggests sediment structure produce by a bidirectional flow, which is associated with tidal setting (LePain et al. 2009; Nichols 2009).

(iv) *Hummocky cross-stratified sandstone sub-facies* is characterised by wavy to swaley cross-stratified medium to coarse-grained sandstone with visible thin conglomeritic lag. Dott and Bourgeois (1982) and Walker and Plint (1992) interpreted this sub-facies as deposits formed by re-deposition of fine sand delivered offshore by flooding rivers and/or scour of the shoreface by large storms, below normal fair-weather wave base.

Miall (1978, 1996), Kendall (2005), Nichols (2009) and LePain et al. (2009) suggested that cross-bedded sandstone facies are associated with deposits are formed by the migration of straight to sinuous crested dunes.

Conglomeritic sandstone facies is a minor component observed only within the stratigraphic succession of Ozara–Ukwu, Afikpo and Ngodo Hill outcrops. This facies although occurring in minor form appears to be common in the southern part and rate in the northern part study area. This facies consists of brown to reddish-brown, pebbly conglomerate with a bed thickness of varying from 0.5 to 4 m (Fig. 4.8h, j). Sandstone and pebbly sandstone units of this facies were partly bioturbated. Evident on the rock surfaces were *Thallassinoides* and *Skolithos* burrows, which indicated that the rock units of this facies were probably lag deposits or transgressive lag above firm ground in a shallow sandy shoreline (MacEachern and Pemberton 1992; Miall 1996; Dim et al. 2016).

4.1.2 Facies Associations

Facies association depicts the seaward or landward progressive migration of depositional environment through time (Walker and Plint 1992). Identified facies occurring in vertical succession were classified into genetically related facies associations. Key attributes such as vertical stacked facies succession, lateral facies transition and sedimentary structures and mineralogy were used to group the facies into five facies association. These observed attributes were utilized in reconstructing the paleo-depositional environment for the various facies of the facies association unravelling their spatial distribution (see Figs. 5.1 and 5.2).

4.1.2.1   Oxygenated Offshore/Open Shelf Deposits Facies Association (FA) 1

Oxygenated offshore/open shelf deposits facies association is common in the outcrops of the central and southern parts of the study area. This facies association consists of mottled shales of the partly laminated silty shale and sub-facies of the mudstone facies. The mottled appearance suggest deposition in well-oxygenated setting in which the original stratification has be destroyed by organisms (Pemberton et al. 1992). Component facies of these facies association dominant the upper section of the deepening upward cycles or fining/thinning upward successions. Boggs (2001) further suggested that the interlaminated sand and shale component of the silty shale facies of this facies association are signature of a low-energy setting deposits, probably below fair-weather and storm wave base (see Fig. 5.2).

4.1.2.2   Storm Wave-Dominated Shelf Deposits Facies Association (FA) 2

The storm wave-dominated shelf deposits facies association consists of the medium–coarse-grained sandstone facies, hummocky cross-bedded sandstone sub-lithofacies, bioturbated sandstone and siltstone facies, plane-parallel laminated sandstone facies. Component facies of these facies association dominant the basal part of the deepening upward cycles or fining/thinning upward successions and upper part of shoaling upward cycles or coarsening/thickening upward successions. The association of storm wave-dominated shelf and oxygenated offshore/open shelf deposits is an indication that both facies association occurred side by side. These are storm wave-dominated shelf deposits. The presence of storm-generated features such as swales and hummocky cross-stratified sandstone indicates that storms were significant in shaping the Eze-Aku Formation shelf (Einsele 2000; Karim 2007). Walker (1985) and Pattison et al. (2007) suggested that the sands were deposited probably during storm events generating turbidity currents on the shelf. Wonham et al. (2014) also suggested that the dominance of quartz and relatively low feldspar of the medium to coarse-grained component of this storm wave-dominated shelf facies association reflects reworking of the shoreface and re-deposited as turbidites on the shelf. The associated bioturbated sandstone and siltstone facies reflects deposits within zones below fair-weather wave base and above storm wave base. Weise (1980) and Walker and Plint (1992) describes these as Blower shoreface and Boffshore transition zones respectively (see Fig. 5.2).

4.1.2.3   Shoreface Deposits Facies Association (FA) 3

The shoreface deposits facies association comprises planar and trough cross-bedded sandstone subfacies, plane/parallel and rippled laminated sandstone facies. The component facies of the shoreface deposits facies association were seen occurring within the upper part of a shoaling upward cycles or coarsening/thickening upward facies successions. This coarsening upward succession is an indication of increasing energy level associated with progradational packages (Ekwenye et al. 2014). The shoreface deposits facies association overlies the mudstone facies of oxygenated offshore/open shelf deposits and hummocky cross-stratified sandstones of the storm wave-dominated shelf deposits. The presence of wave-rippled laminated structure that characterize the sandstone units suggests reworked deposits typical of upper shoreface setting, above fair-weather wave base (Gowland 1996).

Furthermore, the prominent vertical burrows, the planar and trough cross-bedded sandstone sub-facies, plane/parallel and rippled laminated sandstone facies are interpreted to be the result of vigorous current and wave activity, which are typical upper shoreface zone of intermediate to high-energy wave coasts (Howard and Reineck 1979, 1981). Associated biotubated sandstone and siltstone facies, and fine sandstone facies probably depicts deposits of the lower shoreface environment.

4.1.2.4 Tidal Inlet/Lagoonal Deposits Facies Association (FA) 4

The tidal inlet/lagoonal deposits facies association is common over relatively narrow stratigraphic thickness straddling the transition from open marine to alluvial environments in the Afikpo outcrop belt (see Fig. 5.2). This facies association consists of herringbone cross-bedded sandstone and interbedded fine sandstone—siltstone/shale facies, associated with planar cross-bedded and trough cross-bedded subfacies. These facies are interpreted to be bar deposits within tidal inlets. Herringbone cross-stratification indicates reversing current condition (LePain et al. 2009). The presence of muddy drapes on foresets and heterolithic units reflect slack water condition during low-tide and tidally influenced depositional setting (LePain et al. 2009; Hassan et al. 2013). Components of the tide-influenced depositional environments are broken up and washed shell fragments and fossiliferous lime-mud units of the bioclastic limestone facies. These are thought to be patch reef deposits formed in transgressive shallow water.

4.1.2.5 Fluvial Channel Deposits Facies Association (FA) 5

The fluvial channel deposits facies association is common in marginal-marine and continental strata of the Afikpo outcrop. This facies association are seen occurring at the lower part of a deepening upward cycle or fining/thinning upward facies successions. The component facies of this facies association are the muddy conglomeritic sandstone that are characterized by erosional lower contact or marks the channel base. Other minor component facies such as plane-parallel and ripple-laminated sandstone sub-facies suggests channelized unidirectional flows. Howard and Reineck (1979, 1981) interpreted the presence of relatively coarse-grained material indication of waxing fluvial currents, typical of fluvial channels settings.

References

Banerjee I (1980) A subtidal bar model for the Eze-Aku Sandstones, Nigeria. J Sediment Geol 30:133–147

Blatt H, Middleton GV, Murray RC (1980) Origin of sedimentary rocks, 2nd edn. Prentice-Hall, Englewood Cliffs, NJ, 782 p

Boggs SJ (2001) Principles of sedimentology and stratigraphy. Prentice-Hall, Englewood Cliffs, NJ, p 261

Boggs S (2006) Principles of sedimentology and stratigraphy, 4th ed. Pearson Prentice Hall, Upper Saddle River, NJ, 662 p

Collinson JD, Thompson DB (1989) Sedimentary structures, 2nd edn. London, Unwin Hyman, 207 p

Dim CIP, Okwara IC, Mode AW, Onuoha KM (2016) Lithofacies and environments of deposition within the middle–upper cretaceous successions of southeastern Nigeria. Arab J Geosci 9:447. https://doi.org/10.1007/s12517-016-2486-7

Dott RH, Bourgeois J (1982) Hummocky stratification: significance of its variable bedding sequences. Geol Soc Am Bull 93: 663–680

Dunham RJ (1962) Classification of carbonate rock according to depositional texture. In: WE Ham (ed) Classification of carbonate rocks. American Association of Petroleum Geologists Memoir, pp 108–121

Einsele G (2000) Sedimentary basin: evolution, facies and sediment budget, 2nd edn. Springer, 792 p

Ekwenye OC, Nichols GJ, Collinson M, Nwajide SC, Obi GC (2014) A paleogeographic model for the sandstone members of the Imo Shale, southeastern Nigeria. J Afr Earth Sci 96:190–211

Frey RW, Pemberton SG, Saunders TDA (1990) Ichnofacies and bathymetry: a passive relationship. J Paleontol 64:155

Gowland S (1996) Facies characteristics and depositional models of highly bioturbated shallow marine siliciclastic strata: an example from the Fulmar Formation (Late Jurassic), UK Central Graben. Geol Soc Lond Spec Publ 114(10):185–214

Harms JC (1979) Primary sedimentary structures. Ann Rev Earth Planet Sci 7:227–248

Harms JC, Southard JB, Spearing DR, Walker RG (1975) Depositional environments as interpreted from primary sedimentary structures and stratification sequences. SEPM Short Course No. 2, 161 p

Harms JC, Southard JB, Walker RG (1982) Structures and sequences in clastic rocks. Tulsa, OK, Society of Economic Paleontologists and Mineralogists Short Course No. 9, 394 p

Hassan MHA, Johnson HD, Allison PA, Abdullah WH (2013) Sedimentology and stratigraphic development of the upper Nyalau Formation (Early Miocene), Sarawak, Malaysia: a mixed wave-and tide-influenced coastal system. J Asian Earth Sci 76:301–311

Howard JD, Reineck HE (1979) Sedimentary structures of a "high energy" beach to offshore sequence Ventura Port Hueneme area, California, U.S.A. (abstr.). Am Assoc Pet Geol Bull 63:468–469

Howard JD, Reineck HE (1981) Depositional facies of high-energy beach-to offshore sequence: comparison with low-energy sequence. Am Assoc Petrol Geol Bull 65:807–830

Karim KH (2007) Possible effect of storm on sediments of upper Cretaceous Foreland Basin: a case study for Tempestite in Tanjero Formation, Sulaimanyia Area, NE-Iraq. Iraq J Earth Sci 7(2): 1–10

Kendall CG (2005) Introduction to sedimentary facies, elements, hierarchy and architecture: a key to determining depositional setting. University of South Carolina. Spring, USC Sequence Stratigraphy Web Site. http://strata.geol.sc.edu

LePain DL, McCarthy PJ, Kirkham R (2009) Sedimentology, stacking patterns, and depositional systems in the Middle Albian–Cenomanian Nanushuk Formation in Outcrop, Central North Slope, Alaska. Report of Investigations 2009-1, Division of Geological & Geophysical Surveys, pp 1–86

MacEachern JA, Pemberton SG (1992) Ichnological aspects of Cretaceous shoreface successions and shoreface variability in the western interior seaway of North America. In: Pemberton SG (ed) Applications of ichnology to petroleum exploration. SEPM Core Workshop No. 17, pp 57–84

MacEachern JA, Pemberton SG (1994) Ichnological aspects of incised valley fill systems from the Viking Formation of the Western Canada Sedimentary Basin, Alberta, Canada. In: Boyd R, Zaitlin BA, Dalrymple R (eds) Incised valley systems—origin and sedimentary sequences. Society of Economic Paleontologists and Mineralogists, Special Publication, vol 51, pp 129–157

Miall AD (1978) Lithofacies types and vertical profile models in braided river deposits: a summary. In: Miall AD (ed) Fluvial Sedimentology Memoir 5, Canadian Society of Petroleum Geologists, pp 597–604

Miall AD (1996) The geology of fluvial deposits: sedimentary facies, basin analysis, and petroleum geology, 1st edn. Springer-Verlag, Berlin, p 582

Miall AD (2000) Principles of sedimentary basin analysis, 3rd edn. Springer-Verlag, Berlin, Heidelberg, 616 p

Nichols G (2009) Sedimentology and stratigraphy, 2nd edn. Wiley-Blackwell Publications, 419 p

Pattison SAJ, Ainsworth RB, Hoffman TA (2007) Evidence of a cross-shelf transport of fine-grained shelf channels in the Campanian Aberdeen Member, Book Cliffs, Utah, USA. Sedimentology 24:1033–1063

Pemberton SG, MacEachern JA, Frey RW (1992) Trace fossil facies models: environmental and allostratigraphic significance. In: Walker RG, James N (eds) Facies models: response to sea level change. Geological Association of Canada, St. John's, Newfoundland, pp 47–72

Reyment RA (1965) Aspects of the geology of Nigeria: the stratigraphy of the Cretaceous and Cenozoic deposits. Ibadan University Press, Ibadan, p 145

Spring D, Hansen OP (1998) The influence of platform morphology and sea level on the development of a carbonate sequence: the Harash Formation, Eastern Sirte Basin, Libya. In: MacGregor DS, Moody RTJ, Clark-Lowes DD (eds) Petroleum geology of North Africa. Geological Society of London, London, Special Publication, No. 132, pp 335–353

Walker RG (1985) Geological evidence for storm transportation and deposition on ancient shelves. In: Eds RW, Tillman DJP, Swift (eds) Walker Shelf Sands and Sandstone Reservoirs. SEPM Short Course Notes, vol 13, pp 243–302

Walker RG, Plint AG (1992) Wave and storm-dominated shallow marine systems. In: Walker RG, James NP (eds) Facies models, response to sea level change. Geological Association of Canada, St. John's, New Foundland, Canada, pp 219–238

Weise BR (1980) Wave-dominated delta systems of the Upper Cretaceous San Miguel Formation, Maverick Basin, South Texas. Bureau of Economic Geology, University of Texas at Austin Report Inv (107), p 40

Wonham J, Rodwell I, Lein-Mathisen T, Thomas M (2014) Tectonic control on sedimentation, erosion and redeposition of Upper Jurassic sandstones, Central Graben, North Sea. From Depositional Systems to Sedimentary Successions on the Norwegian Continental Margin, pp 473–512

Chapter 5
Stratigraphic Evolution, Provenance and Paleo-Depositional Reconstruction of Facies

5.1 Facies Depositional Model

5.1.1 Evolution and Provenance of Stratigraphic Succession

Evidence from field mapping and geological studies carried out in this work has shown that the stratigraphic successions were affected by several factors, which had influence on their depositional environment. The deposition of the sediment package of the Eze-Aku Formation and associated Amaseri Sandstones of the Southern Benue Trough (constituting the oldest formation in the area) occurred during the Turonian through Coniancian age (Simpsons 1954; Amajor 1987). The stratigraphic succession of the Eze-Aku Formation from outcrop studies reveal thick shales interstratified with sandstone units (Figs. 4.2 and 5.1). This is an indication that sedimentation occurred at various episodes of relative sea-level changes namely; a transgressive episodes that deposited shales during sea-level rise and a regressive episodes that brought in sandstones/siltstones during se- level fall (Dim et al. 2016).

The subsequent mid-Santonian inversion that affected sedimentation left several imprints of tectonic and magmatic activities observed on stratigraphic successions of the study area and the southern Benue Trough, which spans from Cenomanian through the Campanian in age (Olade 1975; Benkhelil 1989). The resulting structural features were folded, tilted and fractured rock units, and igneous rocks emplacement evident in the area. The associated drainage and basin reorganisation led to the establishment of a fluvial system. The fluvial system shed sediment from the uplifted units into the Campanian (Afikpo Formation) Anambra Basin that overlies the Eze-Aku Formation, while the relatively undisturbed units (Afikpo Formation) with low stratigraphic dips observed in the southern part of the study area suggest deposition after the Santonian deformation stage (Hoque and Nwajide 1985; Murat 1972; Petters 1991; Ojoh 1992).

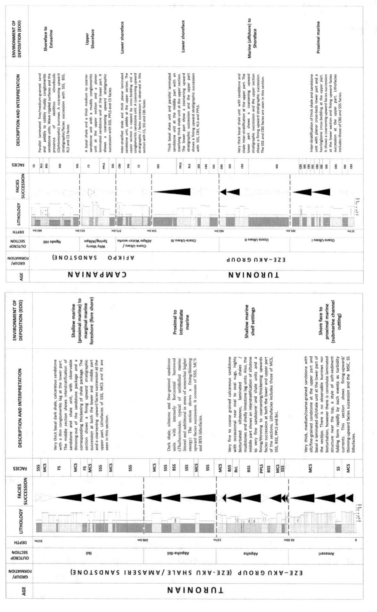

Fig. 5.1 Stacked sedimentologic log of the outcrop sections showing lithofacies, facies succession and their environments of deposition implication (after Dim et al. 2016)

Results from thin section analysis of sandstone units reveal that they are texturally immature arkosic sandstones, with poor to moderate sorting, angular to sub-angular mineral grains, and high feldspar content of up to 25%. The abundance of the little altered feldspars suggests high feldspar content from parent rock that have been transported for a relatively short distance, indicative of a near-source provenance probably derived from the proximal south-eastern basement granites in the Oban Massif (Hoque and Nwajide 1985; Dim et al. 2016).

5.1.2 Facies Distribution and Paleo-Depositional Reconstruction

Results from the analysis of the ten identified lithofacies show that the silty shale sub-facies of the mudstone facies, medium to coarse-grained sandstone, fine sandstone, bioclastic limestone, bioturbated siltstone facies, and trough cross-bedded sandstone sub-facies characterised the outcropping stratigraphic succession of the northern through the central parts of the study area. The shale sub-facies of the mudstone facies, bioturbated sandstone, plane-parallel laminated sandstone, cross-bedded sandstone and conglomeritic sandstone facies characterise the outcropping stratigraphic succession of the southern part of the study area (Table 3.1).

The two facies cycles or successions recognized on the vertical stacked measured/logged stratigraphic succession reveal that the deepening upward cycles or finning/thinning upward facies successions that depicts transgressive episode, associated with retrograding mudstone deposits, dominates the northern through central part of the area. Sediment of this cycle represent deposition in low energy environment. This is contrary to the southern part of the area, which shows predominance of shoaling upward cycles or coarsening/thickening upward facies successions that depicts regressive episode, associated with prograding sandstone deposits (Figs. 4.4 and 5.1; Dim et al. 2016). Sediment of this cycle represents deposited in a shallow marine. However, these facies cycles or successions distributions contradicts the model developed by Banerjee (1980), that reported that the area is characterized by majorly coarsening upward succession, typical of sub-tidal sand bars.

The outcome of classifying of these lithofacies into five facies association enable the reconstruction of the paleo-depositional environments and understanding of the prevailing condition at the time sedimentation. The stratigraphic successions that were formed in distal offshore zone comprises the silty-shale and shale sub-facies of mudstone facies suggest deposition in within oxygenated offshore/open shelf typical of low energy setting, below fair-weather and storm wave base (Fig. 5.2; Banerjee 1980; Pemberton et al. 1992; Boggs 2001; Dim et al. 2016). In the proximal offshore zone are the deposits of medium-coarse-grained sandstone facies, hummocky cross-bedded sandstone sub-facies, bioturbated siltstone and sandstone facies and planar-parallel-laminated sandstone that represents deposition within the storm wave-dominated shelf deposits (Fig. 5.2). Although the works of Okoro and

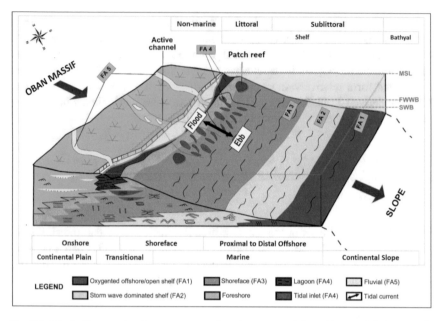

Fig. 5.2 Block diagram showing reconstructed environments of deposition model showing the distribution of facies association of various lithofacies delineated in the Afikpo area and environs (after Dim et al. 2016)

Igwe (2014) suggested a deep-water depositional setting for storm wave-dominated shelf facies association, based on the absence of sedimentary structures of storm origin, the present work, however, reveals the occurrence of hummocky and swaley structures (Fig. 4e; Dim et al. 2016). The abundance of *Planolites* of the *Cruziana* ichnofacies associated with the bioturbated siltstone and sandstone facies suggests deposits of marine setting. The shoreface zone is characterized by planar, trough cross-bedded sandstone sub-facies and wave rippled-laminated sandstone facies that constitute deposits of upper and lower shoreface, which accumulated above fair-weather wave base. The presence of *Ophiomorpha* burrows of the *Skolithos* ichnofacies associated with sandstone facies reflects deposits of the upper shoreface environment.

The transitional zone is made up of stratigraphic succession characterized by herringbone cross-bedded sandstones sub-facies and bioclastic limestone facies that were deposits of transgressive shallow waters and formed within the tidal inlet/lagoonal setting in the shelf environment (Dim et al. 2016). This agrees with work of Hoque (1976), which reported the Eze–Aku Formation contains scattered lenses and layers of bioclastic limestone. The presence of *Thallassinoides* of the *Glossifungites* ichnofacies found in unlithified marine littoral and sub-littoral surfaces suggests moderate-energy settings (Pemberton et al. 1992; Dim et al. 2016). The continental or non-marine zone comprises stratigraphic succession of conglomeritic sandstone (CS) facies, associated cross-bedded sandstone sub-facies, plane/parallel laminated sandstone and rippled-laminated sandstone facies that reflects deposits of fluvial channel. Generally, the analysis of the facies of the outcropping stratigraphic succession of Afikpo area suggests that sedimentation occurred across non-marine through the distal marine depositional setting (Fig. 5.2).

References

Amajor LC (1987) The Eze-Aku Sandstone ridge (Turonian) of southeastern Nigeria: a re-interpretation of their depositional origin. J Min Geol 23:17–26

Banerjee I (1980) A subtidal bar model for the Eze-Aku Sandstones, Nigeria. J Sediment Geol 30:133–147

Benkhelil J (1989) The origin and evolution of the Cretaceous Benue Trough, Nigeria. J Afr Earth Sc 8:251–282

Boggs SJ (2001) Principles of sedimentology and stratigraphy. Prentice-Hall, Englewood Cliffs, NJ, p 261

Dim CIP, Okwara IC, Mode AW, Onuoha KM (2016) Lithofacies and environments of deposition within the middle–upper cretaceous successions of southeastern Nigeria. Arab J Geosci 9:447. https://doi.org/10.1007/s12517-016-2486-7

Hoque M (1976) Significance of textural and petrographic attributes of several Cretaceous sandstones, southern Nigeria. J Geol Soc India 17:514–521

Hoque M, Nwajide CS (1985) Tectono-sedimentological evolution of an enlongate intracratonic basin (aulacogen): the case of the Benue Trough of Nigeria. J Min Geol 21:19–26

Murat RC (1972) Stratigraphy and palaeogeography of the Cretaceous and Lower Tertiary in southern Nigeria. In: Dessauvagie FJ, Whiteman AJ (eds) African geology. University of Ibadan Press, Ibadan, Nigeria, pp 251–266

Ojoh KA (1992) The southern part of the Benue Trough (Nigeria) Cretaceous stratigraphy, basin analysis, paleo-oceanography and geodynamic evolution in the Equatorial domain of the South Atlantic. Niger Assoc of Pet Explorationists' Bull 7(2):131–152

Okoro AU, Igwe EO (2014) Lithofacies and depositional environment of the Amasiri Sandstone, Southern Benue Trough, Nigeria. J Afr Earth Sci 100:179–190

Olade MA (1975) Evolution of Nigerian Benue Trough (aulacogen): a tectonic model. Geol Mag 112:575–583

Pemberton SG, MacEachern JA, Frey RW (1992) Trace fossil facies models: environmental and allostratigraphic significance. In: Walker RG, James N (eds) Facies models: response to sea level change. Geological Association of Canada, St. John's, Newfoundland, pp 47–72

Petters SW (1991) Regional geology of Africa. Springer Verlag, Berlin, Heidelberg, p 722

Simpsons A (1954) The geology of parts of Onitsha, Owerri, and the Nigerian coal fields. Geol Surv Niger Bull 24:121

Conclusions

An integrated study utilizing information obtained from geologic field mapping and detailed measured/logged stratigraphic sections was carried out across the sandstone ridge system of Afikpo area and its environs within the southeastern part of Nigeria. This study incorporates descriptions from ten outcrop section and interpretive results from facies and petrographic analysis in improving our understanding of outcropping lithofacies, their distribution and paleo-depositional environments.

Studies indicate that the stratigraphic units of the area belongs to the Mid–Upper Cretaceous succession of Eze-Aku Formation (Southern Benue Trough) and Afikpo Formation (part of the basal sequence of the Anambra Basin). Field observation revealed that outcrop sections were characterized by series of structurally deformed stratigraphic packages, which were caused by the Santonian inversion that affected the area. Tilted sandstone beds associated with the deformation formed alternating parallel to sub-parallel ridges with the shales that dominates the lowlands. Major structural features evident in the area were igneous emplacements (dolerite sill), anticlines, fold limbs, deformational bands, faults and joints. These features were mainly post-depositional structures.

Facies analytical studies reveal that the area is characterized by ten outcropping lithofacies. The distribution of these lithofacies were such that the medium–coarse sandstone and mudstone facies predominates and were abundant in occurrence across the area. The fine sandstone, bioturbated siltstone and planer-parallel-laminated sandstone facies were common relative to the bioclastic limestone, bioturbated sandstone, ripple-laminated sandstone, cross-bedded sandstone and conglomeritic sandstone facies, which occurred in minor forms. In addition, petrographic (thin section) study shows that the medium to coarse-grained sandstones facies were arkosic and texturally immature with relatively high feldspar content, which is an indicative of a near-source provenance, probably the adjoining Oban Massif.

Measured/logged profiles from stratigraphic sections unravelled two stacking pattern that suggested two facies succession. The fining-upward/thinning-upward stacking pattern depicts a deepening upward facies succession whereas the

C. I. P. Dim et al., *Facies Analysis and Interpretation in Southeastern Nigeria's Inland Basins*, SpringerBriefs in Earth Sciences, https://doi.org/10.1007/978-3-030-68188-3

coarsening-upward/thickening-upward stacking pattern depicts a shoaling upward facies succession. Furthermore, studies revealed that the identified lithofacies belong to five facies associations. These facies associations were deposited within the oxygenated offshore/open shelf, storm wave-dominated shelf, shorefaces, tidal inlet/lagoonal and fluvial channels settings. Hence, reconstructed depositional model shows stratigraphic packages deposited within environments that spanned from shallow marine through proximal and distal marine.